Climate Change and its Causes, Effects and Predictions Series

LAND USE AND CLIMATE CHANGE

Climate Change and its Causes, Effects and Predictions Series

LAND USE AND CLIMATE CHANGE

SURESH C. RAI

Nova Science Publishers, Inc.
New York

For permission to use material from this book please contact us:
Telephone 631-231-7269; Fax 631-231-8175
Web Site: http://www.novapublishers.com

NOTICE TO THE READER

The Publisher has taken reasonable care in the preparation of this book, but makes no expressed or implied warranty of any kind and assumes no responsibility for any errors or omissions. No liability is assumed for incidental or consequential damages in connection with or arising out of information contained in this book. The Publisher shall not be liable for any special, consequential, or exemplary damages resulting, in whole or in part, from the readers' use of, or reliance upon, this material. Any parts of this book based on government reports are so indicated and copyright is claimed for those parts to the extent applicable to compilations of such works.

Independent verification should be sought for any data, advice or recommendations contained in this book. In addition, no responsibility is assumed by the publisher for any injury and/or damage to persons or property arising from any methods, products, instructions, ideas or otherwise contained in this publication.

This publication is designed to provide accurate and authoritative information with regard to the subject matter covered herein. It is sold with the clear understanding that the Publisher is not engaged in rendering legal or any other professional services. If legal or any other expert assistance is required, the services of a competent person should be sought. FROM A DECLARATION OF PARTICIPANTS JOINTLY ADOPTED BY A COMMITTEE OF THE AMERICAN BAR ASSOCIATION AND A COMMITTEE OF PUBLISHERS.

LIBRARY OF CONGRESS CATALOGING-IN-PUBLICATION DATA

Rai, Suresh Chand, 1958-
 Land use and climate change / Suresh C. Rai.
    p. cm.
 Includes bibliographical references and index.
 ISBN 978-1-60741-362-2 (hardcover)
 1. Land use. 2. Climatic changes. I. Title. HD111.R24 2009
 551.6--dc22   2009008026

Published by Nova Science Publishers, Inc. ✦ *New York*

CONTENTS

PREFACE

During recent past, large scale conversions of forests to other land-uses in developed and developing countries in response to high population growth has been observed. As a consequence, this has disrupted the hydrological cycle and a great change in climate is envisaged. The land-use change from forest to the other usage has been quite conspicuous in the last few decades in the Indian Himalayan region. Land-use/cover change is emerging as a central issue within the community concerned with global environment. This concern is driven by the contributions that land-use/cover transformations make a wide variety of changes. The recent interest in land-use change has generated effort to understand the interrelationship between land-use/cover and climate change. Most of the global environmental problems fall within the gambit of land-use/cover change and require a thorough understanding of its causes and consequences. Understanding of land-use/cover and climate change is important for the sustainable development and to evolve strategies for the mitigation of climate change at local, regional and global levels. This work is basically designed to familiarize the graduate, post graduate and research students with the basic knowledge of land-use/cover and its impact on climate.

The present study has been divided into five chapters dealing with varied but interrelated aspects. Chapter one deals with an introductory outline of the land-use and climate change, IGBP initiatives on land-use/cover and climate challenges, the Kyoto Protocol and India's environmental issues. Since not much study has been carried out on the topic in the Indian context, more attention is paid to review studies carried out in other countries. Chapter two is devoted to describe the land-use pattern, land-use/cover change detection on global, regional and local level. The chapter on biogeochemical cycles (third chapter) deals with carbon, nitrogen, phosphorous, and sulfur cycles as influenced by human activities. It is followed by hydrological analyses (chapter four). The global climate change has been discussed in chapter fifth.

The author would like to thank individuals, organizations, and academic institutions whose research publications, technical reports, annual reports and books provided the much needed information for developing the text, tables and figures. Author thankfully acknowledges all the original sources of information and the concerned publishers.

The author wish to acknowledge all those who have helped and provided useful input into this work such as Dr. A.K. Saha for his unending support, right suggestions, and constant encouragement throughout my work. Despite his busy schedule, he never failed to clear my confusions and showed me the right direction. I am thankful to the Head, Department of Geography, Delhi School of Economics, University of Delhi for providing the necessary facilities. I am highly grateful to Prof. J.S. Singh, FNA, Department of Botany, Banaras Hindu University, Professor S.K. Aggarwal (Retd.), Department of Geography, Delhi School of Economics, University of Delhi and Prof. K.M.M. Dakshini (Retd.), Department of Botany, Delhi University for their visionary ideas and time to time suggestions.

I am especially indebted to my wife Smt Kulwanti Rai for her patience and support.

ABOUT THE AUTHOR

Dr. Suresh C. Rai, M. Sc., Ph.D. (Banaras Hindu University) is teaching physical geography at the Department of Geography, Delhi School of Economics, University of Delhi, Delhi-110007 since 2005. Before joining University of Delhi he worked as Senior Scientist in G.B. Pant Institute of Himalayan Environment and Development (An Autonomous Institute under Ministry of Environment and Forests, Government of India) at its Sikkim and North-East Unit and conducted research work on Integrated Watershed Management, Carbon Dynamics and Eco-tourism aspects. He was recipient of Honorable Mention Best Paper Award, 1998 from Soil and Water Conservation Society, USA. His principal area of research is watershed management, mountain hydrology and ecotourism and biodiversity conservation of Himalaya. He has published more than 60 research papers in leading national and international journals and authored/edited 3 books.

Chapter 1

INTRODUCTION

1.1. BACKGROUND

Deforestation, urban sprawl, agriculture and other human activities have substantially altered and fragmented our landscape. Such disturbances of the land can change the global atmospheric concentration of carbon dioxide, the principal heat-trapping gas, as well as affect local, regional and global climate by changing the energy balance on earth's surface. Change and variability in land-use by humans and the resulting alterations in surface features are major but poorly recognized drivers of long-term global climate patterns (Pielke Sr., 2005). Conversion of land from forest to other usage is a threat to land-use sustenance and contributes negatively to climate change. The most important anthropogenic influences on climate are the emission of greenhouse gases and changes in land-use, such as urbanization and agriculture. Land-use/cover changes have caused a significant release of CO_2 to the atmosphere from the terrestrial biota and soils. Soil is important for sequestering atmospheric CO_2 (Box 1.1). Feddema *et al.,* (2005) reported that future land-use/cover will continue to be an important influence on climate for the next century. Although the impact of land-use/cover on the atmospheric concentration of carbon dioxide and methane, and on the global average surface albedo, have been included in international climate change assessment, the role of land-use/cover change and variability in altering regional temperatures, precipitation, vegetation, and other climate variables has been mostly ignored (Pielke Sr., 2005).

The importance of land-use/cover change and variability should not be a surprise. According to NASA reports "Scientists estimate that between one-third and one-half of our planet's land surfaces have been transformed by human development (NASA news feature, 2005). A large body of research has documented the major role of land-use/cover change and variability in the climate systems (Pielke Sr., 2005; Kalnay & Cai, 2003; Feddema, *et al.*, 2005; Galo *et al.*, 1996). Land-use/cover change plays a pivotal role in global environmental change. It contributes significantly to earth-atmosphere interactions and biodiversity loss, is a major factor in sustainable development and human responses to global change, and is important to integrated modeling and assessment of environmental issues in general (Meyer & Turner, 1994).

Box 1.1. Carbon Sequestration

book

Carbon sequestration refers to the provision of long-term storage of carbon in the terrestrial biosphere, underground, or the oceans so that the buildup of carbon dioxide (the principle greenhouse gas) concentration in the atmosphere will reduce or slow.

The human impact on the global environment has received wide attention in the past. Marsh (1864) recognized the deleterious consequences of human activities on the earth's landscape. More recently, Thomas (1956) lent further credence to the notion that one of the most obvious global changes in the last three centuries has been the direct human modification and conversion of land-cover. Land-use changes are cumulatively transforming land-cover at an accelerating pace (Turner *et al.* 1994). These changes in terrestrial ecosystems are closely linked with the issue of the sustainability of socio-economic development since they affect essential parts of our natural capital, such as climate, soils, vegetation, water resources and biodiversity (Mather & Sodsyuk, 1991). Land-cover transformation is significant to a range of themes and issues central to the study of global environmental change/climate change. The alterations and its effect on the surface of the earth hold major implications for sustainable development and livelihood systems and also contribute to changes in the biogeochemical cycles of the elements affecting the atmospheric levels of greenhouse and other trace gases.

The land-use change from forest to other usage has been quite conspicuous in the last few decades in the developing countries. The consequences of land-use transformation from forest to agriculture, in the developing countries of the tropics has aroused international concern for human poverty, loss of plant and animal species, erosion of landscape, siltation of water courses, and flooding (Turner *et al.*, 1990). The reduction of original forest cover already amounts to at least 21% in Asia and Australia (Jackson, 1983; Rubinoff, 1983). These changes have contributed to manifold and massive alterations of fundamental biochemical cycles (e.g., carbon cycle) and have been a major contributory source to increased CO_2 concentration in the atmosphere in the region (Clark, 1982).

The historical conversion of natural systems to agriculture and other human uses of the land have resulted in a net release of CO_2 to the atmosphere (Houghton et al., 1985). The land has been a source of carbon to the atmosphere since about 1860 when agricultural expansion began, and until the end of the 1970s more carbon came from terrestrial ecosystems than from fossil fuel combustion (Houghton, et al., 1983). Schlesinger (1995) estimated that annual net release of carbon from agricultural activities is about 0.8 Pg yr^{-1}, or about 14% of the current fossil fuel emissions. Total change in land-use for different eco-regions is estimated at 17×10^6 km^2 (Williams, 1994). Major changes in land-use occurred in the forests (7.0×10^6 km^2 or 39.8%) and grassland (6.6×10^6 km^2 or 37.5%) ecosystems accounting for 77.3% of the total land-use change. A significant proportion of this change has occurred since the early 1800s. Simulations of tropical deforestation and potential future human land-cover impacts project a warming of 1-2^{0}C in deforested areas (Feddema et al., 2005). Land-cover impacts on global climate can be divided into two major categories: (i) biogeochemical, and (ii) biogeophysical. Biogeochemical processes affect climate by altering the rate of biogeochemical cycles, thereby changing the chemical composition of the atmosphere. To some extent, these emissions are included in the IPCC climate change assessments (Houghton et al., 2001). Biogeophysical processes directly affect the physical parameters that determine the absorption and disposition of energy at earth's surface. Albedo, or the reflective properties of earth's surface, alters the absorption rate of solar radiation and hence energy availability at earth's surface (Feddema et al., 2005).

The importance of land-use/cover change and its role in climate change is duly recognized by the International Geosphere Biosphere Programme (IGBP) and the International Human Dimension Programme (IHDP) on Global Environmental Change (GEC). The IGBP also has a focus on these issues through the programme on Biospheric Aspects of Hydrological Cycle (BAHC), Global Change and Terrestrial Ecosystem (GCTE) and Land-Use/Cover Change (LUCC). In order to address the consequences of global change in mountains around the world, an initiative for collaborative research on global change and mountain regions-the Mountain Research Initiative (MRI) - was developed and officially launched in July 2001. It will involve close collaboration between these organizations.

The pace, magnitude and spatial reach of human alterations of the earth's land surface are unprecedented. To understand recent changes and generate scenarios on future modifications of the Earth System, the scientific community needs quantitative, spatially explicit data on how land-cover has been changed by human use over the last few decades and how it will be changed in the future.

1.2. LAND-USE/COVER CHANGE

The terrestrial or land-covers of the earth and changes therein are central to a large number of the biophysical processes of global environmental change. Land-cover is the biophysical state of the earth's surface and immediate subsurface. Changes in land-cover include changes in biotic diversity, actual and potential primary productivity, soil quality and runoff and sedimentation rates (Steffen *et al.*, 1992). Land-covers and changes in them are sources and sinks for most of the material and energy flows that sustain the biosphere and geosphere, including trace gas emissions and the hydrological cycles (BAHC, 1993; Holligan & de Boois, 1993; Matson & Ojima, 1990).

The antiquity of land-cover changes is reflected in their prominence in the early classics of environmental science. George Perkins Marsh's *Man and Nature* (Marsh, 1864) was a monumental assessment of data and theories, many dating back much earlier, on the effects of land-cover changes, particularly deforestation. Thomas (1956) argued that human driving forces play a key role to altered terrestrial ecosystems since, at least, the use of fire to hunt and the advent of plant and animal domestication. Such changes increased dramatically throughout the agricultural phase of history (Wolman & Fournier, 1987), most strikingly in deforestation (Williams, 1990) and the transoceanic movement of species (Crosby, 1986). These changes were of no small consequences, and yet in spatial scale, magnitude, and pace they play in comparison to those produced by modern industrial society. Today, land-cover change of many kinds are global in spatial scale and magnitude and rapid, if variable, in pace, some of them large enough to contribute significantly to changes in global biogeochemical flow.

Over the course of 20th century, humans have emerged as a primary cause of land-cover change around the world (Allen & Barnes, 1985; Turner *et al.*, 1990; Whitby, 1992), an understanding of land-use change is essential to understanding land-cover change (Box 1.2).

The widespread transformation of land is mainly through efforts to provide food, shelter and products for human use. Land-use change is now recognized as an issue that is environmentally significant. The effects of land-use change and management are so significant that collectively they form one of the major environmental changes that are occurring at a global scale (Dale *et al.*, 2000). To understand human-induced change in land-cover, therefore, requires an understanding of its underlying social causes. This is especially true considering that most of the earth's land is already damaged/changed.

Box 1.2. Definition

Land-use: The term denotes the human employment of the land. Land-uses include settlement, cultivation, pasture, rangeland, recreation and so on.

Land-use Change: At any location may involve either a shift to a different use or an intensification of the existing one.

Land-cover: Denotes the physical state of the land. It embraces for example, the quantity and type of surface vegetation, water and earth materials.

Land-cover Change: Falls into two ideal types, conversion and modification. The former is a change from one class of land-cover to another: from grassland to cropland, for example. The latter is a change of condition within a land-cover category, such as the thinning of a forest or a change in its composition.

Source: IGBP Report No. 35.

Global inventories of arable land have started date back at least a century (Ravenstein, 1890) and those of forest resources almost as far (Zon & Sparthawk, 1923). Surveys of global change such as the World Resources Institute reports and the recent volume *The Earth as Transformed by Human Action* (Turner, 1990) assemble much historical and statistical material and outline the broad global and regional trend. A SCOPE Volume on *Land Transformation in Agriculture* (Wolman & Fournier, 1987) covers the principal agricultural impacts on land-cover. Recently, efforts have been made to quantify the nature and extent of land-use/cover changes at a global scale. Richards (1990) estimated that over the last three centuries, the total global area of forests and woodlands diminished by 12×10^6 million km^2 (19%), grasslands and pasture declined by 5.6×10^6 million km^2 (8%), and cropland increased by 12×10^6 million km^2 (466%). Such large changes in land-cover can have important consequences such as significant changes in regional and global climate (Bonan, 1999; Dickinson & Henderson, 1988), modification of the global cycles of carbon, nitrogen and water (Houghton *et al.*, 1983) and increased rates of extinction and biological invasion (Vitousek *et al.*, 1997). Change accelerated globally, in terms of both the conversion of lands to cultivation and the intensification of agriculture on land already cultivated. Despite recent deforestation in parts of the tropics for livestock production, the area of rangeland and pasture has remained virtually the same over the last 300 years (Richards, 1990).

Cropland expansion will undoubtedly continue in the near future, but land-cover modification, through increasing intensification of agriculture, is likely to be of greater importance than further land-cover conversion (Ruttan, 1993). Southgate (1990) provides a simple illustration: rising interest rates or agricultural prices will increase deforestation because they provide an incentive for further clearing. Most of the prime agricultural lands of the world, with the exception of some areas in the tropics, are already cultivated, and major increase in food production are likely to come from yield improvements on these lands through the application of fertilizers, pesticides and herbicides and irrigation. Irrigation of cropland has expanded some 24-fold over the past 300 years, with most of that increase taking place in this century. This practice has increased methane emissions, while the increasing frequency of land tillage world-wide has affected soil carbon (Cole *et al.*, 1989; Rozanou *et al.*, 1990).

Despite the recognition of the magnitude and impact of global scale changes in land-use/cover, there have been relatively few comprehensive studies of these changes. Several continental-to-regional scale land-use data sets have been compiled. For example, Houghton (1990) presents land-use data for nine continental-scale regions of the world. Richards & Flint (1994) have compiled a very comprehensive land-use data base for south and south-east Asia.

In the Indian context, a rapid transformation of land-use has led to environmental degradation and economic deterioration in the Himalaya, where majority of people are living just at or below the subsistence level (Thapa & Weber, 1990). The ecological consequences and the level of degradation of the fragile ecosystems of the Himalaya are well perceived and addressed by many national and international organizations for promoting more effective conservation of natural resources. But it is disheartening that most of the projects/study have proceeded without adequate knowledge of local land practices and environments and, perhaps even more importantly, without an adequate understanding of the capabilities and limitations of the people within them (Stone, 1990).

Himalayan Mountain that holds the largest contiguous tropical to temperate forest in the world is going severe deforestation due to consequences of land-use change (Shah, 1982; Singh *et al.*, 1984; Rai *et al.*, 1994; Rai, 1995). Large-scale deforestation in the region started since 1823 when the British decided to expand the amount of arable land, and by the late 1860s in Central Himalaya cultivated land had more than doubled. During 1840s and 1850s constituted the first era of large-scale uncontrolled deforestation to meet the timber demands of the people

(Trucker, 1983). The commercial exploitation of forests has since continued along with the expansion of agriculture. Shah (1982) argue that continued population growth has led to more farming, and as a result the area under cultivation has increased at a rate of 1.5% yr^{-1} and the cattle population at a rate of 0.18% yr^{-1}. Singh et al., (1984) estimated that only 28.7% of the Indian Central Himalaya is now forested, and that only 4.4% of the area has a forest with greater than 60% crown density. Conditions in neighboring countries are no better. Studies indicate that this will affect the increase of carbon dioxide in the atmosphere, regional hydrology and climate (Singh et al., 1985). Therefore, a better understanding of land-use/cover change is required for the climate change.

1.2.1. Land-use/Cover and Biogeochemical Cycles

Terrestrial ecosystems are important components in the biogeochemical cycle that create many of the sources and sinks of CO_2, CH_4 and N_2O and thereby influence global responses to human-induced emissions of greenhouse gases. Land-cover conversion is an important historical and contemporary component of other forms of global change (IGBP Report No. 35). The historical conversion of natural systems to agriculture and other human uses of the land has resulted in a net release of carbon dioxide to the atmosphere (Houghton et al., 1985; Houghton et al., 1987; Houghton & Skole, 1990; Sharma, 2003), one roughly equivalent to the release from fossil fuel burning over the last 150 years, although the current release of carbon dioxide from land-cover conversion is approximately 30% of fossil fuel combustion (IGBP Report No. 35). Land-cover conversion may have an important influence on regional climatology and hydrology (Shukla et al., 1990).

Both land-use/cover change data are important for determining the biogeochemical cycling of carbon, nitrogen and other elements at regional to global scales. Conversion of tropical forest to pasture seems to be an important factor in trace gas dynamics for years after pasture formation. Land is often converted through biomass burning, which may be an important source of methane, carbon monoxide and other radiatively important trace gases (Crutzen & Andreae, 1990). The atmospheric concentrations of CO_2 and other trace gases are closely linked to each other through their involvement with and interactions in chemical processes in the atmosphere (Prinn, 1994). Land-cover change has an important influence on water and energy balance. Land cover determines surface roughness, albedo, and latent and sensible heat flux, and changes in the distribution of land-cover alter the regional, and possibly global, balance of these fluxes (IGBP Report No. 35).

In Indian context, biomass burning associated with shifting cultivation areas from the north-eastern region is an important source of trace gas emissions in the Southeast Asian region. Data suggested that nearly 112.99 km^2 of the north-eastern region of India affected due to shifting cultivation. Study suggested emissions of 2.063Mt CH$_4$, 17.94 Mt CO, 1.419 Mt N$_2$O, and 51.28 Mt NOx, and 2.643 Mt release of CH$_4$, 3.7204 Mt CO, 0.145 Mt N$_2$O, and 8.477 Mt NOx, respectively, from biomass burning due to shifting cultivation for the year 2000, from the north-eastern region in India (Prasad *et al.*, 2003).

1.2.2. Land-use/Cover and Hydrology

Local hydrology of every river in the world is likely to be affected by climate change in some way. Climate change affects different aspects of local hydrology of river such as timing of water availability and quantity, as well as its quality. Changes in river hydrology will induce risks to water resources facilities that includes flooding, landslides and sedimentation from more intense precipitation events (particularly during the monsoon) and greater unreliability of dry season flows that possesses potential serious risks to water and energy supplies in the lean season.

Land-cover change has an important influence on water and energy balance. Changes in vegetative cover, which mediate the water balance, also influence actual evapotranspiration. Changes in land-cover, therefore, may trigger changes in the hydrological cycle which in turn would have significant implications for land-uses. The impacts of the hydrological cycle caused by land-use/cover changes in the Amazon and Hindu-Kush Himalayan region are not yet adequately assessed. About 70% of the global fluxes are produced in Southern Asia and large Pacific Islands, an area dominated by mountainous source areas, rapidly increasing population, and land cover change (Milliman & Meade, 1983).

The hydrological regime of the Himalayan river catchments is seriously affected by the deforestation of hill slopes. This has caused accelerated erosion particularly in areas where human activities have induced drastic changes in land-use pattern (Singh *et al.*, 1983; Bartarya & Valdiya, 1989; Valdiya & Bartarya, 1991; Rawat & Rawat, 1994; Rai & Sharma, 1998a, b; Sharma *et al.*, 2001). Soil loss and degradation and sediment transport have undoubtedly been increased greatly as a consequence of land-cover change. The recent survey shows that 44% of total land of the Himalayan region is already identified under wastelands.

According to one estimate, nearly 85% of all agricultural land already suffers from severe erosion problems (Shah, 1982). This problem is aggravated by population growth and land transformation especially road-building activities. The present rate of erosion in the catchment's areas of the Himalayan Rivers (100 cm/1000 years) is five times higher than the rate prevailing in the past 40 million years (21 cm/1000 years) (Singh *et al.*, 1984). It can be seen that values of sediment load for the Hindu-Kush Himalayan region exceed the world average by almost two folds (Alford, 1992). Himalayan river increased chemical weathering and associated CO_2 consumption rates could alter the atmospheric CO_2 levels and hence the global climate (Sarin, 2001). Therefore, the relationship between land-use/cover change and soil erosion and hydro-ecological process is an imperative for the Himalayan region.

1.2.3. Land-use/Cover and Climate Change

Among the most obvious consequences of land-use/cover change is the global warming. Over the past few decades, human activities have significantly altered the atmospheric composition, causing a climate change. Land-use change directly affects the exchange of greenhouse gases between terrestrial ecosystems and the atmosphere. The atmospheric concentration of greenhouse gases like methane, nitrous oxide and carbon dioxide continues to increase. The atmospheric CO_2 is currently 353 ppmv and increasing by 0.5% per year (Eriksson, 1991). The atmospheric CH_4 is currently at 1.8 ppmv, and is increasing by about 1% per year, whereas N_2O concentration is 315 ppbv and is increasing by 0.25% per year (Denmead, 1991).

On the Indian sub-continent, temperatures are predicted to increase between 3.5 and 5.5^{0}C by 2100 (IPCC, 2001a) and an even greater increase is predicted for the Tibetan Plateau (Lal, 2002). It is estimated that a 1^{0}C rise in temperature will cause alpine glaciers worldwide to shrink as much as 40% in area and more than 50% in volume as compared to 1850 (IPCC, 2001b). Today, glaciers in the region are retreating, this is compelling evidence of global climate change; if the trend continues, a long-term loss of natural fresh water storage is predicted to be dramatic.

Very few researches have documented the major role of land-use/cover change and variability in the climate system (Pielke Sr. 2005; Kalnay & Cai, 2003). One example of how land-use/cover change affects global climate is the changing spatial and temporal pattern of thunderstorms. Land-use/cover change and variability modify the surface fluxes of heat and water vapor. This alteration

... .s affects the atmospheric boundary layer, and hence the energy available for thunderstorms (Pielke Sr., 2005). Most thunderstorms (by a ratio of about 10 to 1) occur over land and so land-use/cover changes have a greater impact on the climate system.

The conversion of forests to agriculture and other human uses leads to a net release of carbon dioxide to the atmosphere (Likens *et al.*, 1970; Clarke, 1982; Palm *et al.*, 1986; Houghton *et al.*, 1985, 1987, 1990, 2000; Houghton & Skole, 1990; Singh *et al.*, 1991; Wagai *et al.*, 1998; Malhi *et al.*, 1999; Sharma & Rai, 2007). The reduction of original forest cover is one of the most important sources of CO_2 emissions into the atmosphere (Clarke, 1982; Houghton, 1990). Deforestation accounts for substantial release of carbon, one third of which could be due to oxidation of soil carbon in tropics occasioned by changes in land-use pattern (Sanchez *et al.*, 1983; Mann, 1986; Dalal & Meyer, 1986; Bouwman, 1990; Singh *et al.*, 1991; Batjes, 1992). Deforestation not only transfers carbon stocks directly to the atmosphere by combustion, but it also destroys a valuable mechanism for controlling atmospheric CO_2. Historical land-cover conversion by humans may have decreased temperatures by 1^0 to 2^0C in mid-latitude agricultural regions. Simulations of tropical deforestation and potential future human land-cover impacts projects a warming of 1^0 to 2^0C in deforested areas, with possible extra-tropical impacts due to teleconnection processes (Feddema *et al.*, 2005). In the Indian context, the Himalayan Mountains are supposed to be the most sensitive to these effects. Rapid glacial melt can cause serious flood damage in highly populated low lying areas, and as the glaciers recede, large areas that rely on the glacial runoff for water supply could experience severe water shortage (McDowell, 2002). On the much wider perspective, the expected rise in sea level will have profound effects on human sustenance and sustainability of coastal habitats around the world including India (IPCC, 2001).

 Possible adverse consequences of climatic changes resulting from increasing levels of atmospheric CO_2 have drawn attention to the inventory and dynamics of carbon in the biosphere (Chan, 1982; Kellogg, 1982). Carbon exchange between the terrestrial ecosystems and the atmosphere is one of the key processes that need to be assessed in the context of the Kyoto Protocol (IGBP Terrestrial Carbon Working Group, 1998).

1.2.4. Land-use/Cover and Urban Heat Island

Urban heat islands results partly from the physical properties of the urban landscape and partly from the release of heat into the environment by the use of energy for human activities such as heating buildings and powering appliances and vehicles. The global heat flux from this is estimated as 0.03 w m^{-2} (Nakicenovic, 1998). If this energy release were concentrated in cities, which are estimated to cover 0.046% of the earth's surface (Loveland *et al.*, 2000), the mean local heat flux in a city would be 65 wm^{-2}.

Global warming has obtained more and more attention because the global mean surface temperature has increased since the late 19th century. As more than 50% of the human population lives in cities, urbanization has become an important contributor for global warming. Urbanization, the conversion of other types of land to uses associated with growth of populations and economy, is a main type of land-use/ cover change in human history. It has a great impact on climate. By covering with buildings, roads and other impervious surfaces, urban areas generally have higher solar radiation absorption and a greater thermal capacity and conductivity, so that heat is stored during the day and released by night. Therefore, urban areas tend to experience a relatively higher temperature compared with the surrounding rural areas. The temperature difference between the urban and the rural areas are usually modest, averaging less than 1°C, but occasionally rising to several degrees when urban, topographical and meteorological conditions are favourable for the UHI to develop (Mather, 1986). This thermal difference in conjunction with waste heat released from urban houses, transportation and industry; contribute to the development of urban heat island.

Urban Heat Island (UHI) has long been a concern for more than 40 years. Deosthali (2000) found that at night, the core of the city appeared as both heat and moisture islands. The average maximum UHI is weakest in summer and strong in autumn and winter (Kim & Baik, 2002). UHI intensity is related to patterns of land-use/cover changes e.g., the composition of vegetation, water and built-up and their changes. Hence, qualitative studies on the relationship between land-use/cover pattern and LST will help us in land-use planning.

1.3. IGBP INITIATIVES ON LAND-USE/COVER AND CLIMATE CHALLENGE

Environmental change on a global scale became a matter of public concern in the 1960s. Before then, the more widely perceived environmental problem had been urban pollution, which affected human health and the quality of life of so many people, but only on a local scale. The perception that the environment is changing rapidly at the global scale as a result of human actions is relatively recent. In recent times the land-use/cover change is emerging as a central issue within the international communities concerned with global environmental change as it not only has local and regional impacts, but also has important effects at a much larger scale (Richards, 1990). For example, man-made changes in land-use over the last 150 years have contributed about as much carbon dioxide to the atmosphere as has come from fossil fuel combustion (Houghton, 1999). The issue became part of the political agenda, when a meeting of the Heads of Government in Stockholm was organized by the United Nations Environmental Programme, to discuss environmental issues (Appendix-I). Attention was drawn to the acidification of lakes and rivers, global warming caused by trace gases, the destruction of the ozone layer, likely deficits in the supply of food and resources, destruction of rain forest, changes in land-use, biological invasions, and the imbalance in the natural cycles of many of the elements notably carbon and nitrogen (Grace, 2004). By the late 1980s interest in climatic warming had become intense. The Intergovernmental Panel for Climate Change (IPCC) was founded on 1988 to provide expert advice to governments and policy-makers. The inauguration of the International Geosphere-Biosphere Programme (IGBP) in 1992 was an important milestone, enabling international co-ordination of the scientific effort. In 1992 the political leaders of the world met in Rio de Janeiro to set out an agenda to address the environmental, economic and social challenges facing the international community.

Importance of this issue is attested by the emerging International Geosphere-Biosphere Programme and the Human Dimensions Programme's Science agenda on land-use/cover change (IGBP-HDP LUCC) (Turner *et al.*, 1993). There are many other international panels, workshops, and symposia devoted to the topic, i.e. 1991 Global Change Institute on Global Land-use/cover Change of the office of Interdisciplinary Earth Studies (Meyer & Turner, 1994), the 1993 Symposium on 'Land Use and Land Cover in Australia: Living with Global Change', and the 'South East Asian Global Change System for Analysis, Research and Training' (START) programme. This concern is driven by the facts that land transformations bring about a wide variety of global changes- including greenhouse gases and potential global warming, loss of biodiversity, and loss of soil resources and the regional impacts (IGBP, 1998).

The IGBP and IHDP jointly commissioned a core Project Planning Committee/Research Programme Planning Committee for land-use/cover change (CPPC/RPPC LUCC) to create a science/research plan for a jointly sponsored LUCC core project/research programme. Simultaneously with the development of this, the demand for global land-use data base also emerged in the IGBP community. The evaluation of carbon dynamics under such land-use change thus requires a detailed description of activitities both in time and space. Historic and current land-cover and its land-use have to be portrayed and changes have to be adequately monitored (IGBP, 1998). The current availability of comprehensive data sets covering and integrating all these aspects is poor. This is one of the reasons that LUCC, GAIM, GCTE, and PAGES are prioritizing the development of a historic land-cover and land-use database and simulating GCOS/GTOS to develop the necessary observations for monitoring land-use/cover change for future.

IGBP/IHDP-LUCC and IGBP-PAGES came together to take-up the challenge of providing the global change community with historical land-use data sets. PAGES, having participated in the BIOME 6000 projects, have experience with historical reconstructions for 6000 years before present. A new joint PAGES-LUCC initiative, labeled BIOME 300, was created to reconstruct historical land-use/cover data sets for the last 300 years (1700 to 2000) with coarse time slices in the past (50-100 years) and finer time slices in the later periods (10-25 years) (Ramankutty *et al.*, 2001). LUCC is currently recognized as one of critical gaps in our knowledge of the terrestrial carbon cycle which in turn has implications for the rate of greenhouse gas accumulation in the atmosphere and the potential climate change.

The Global change and Mountain Regions Research Initiatives is based on geographical feature- mountain regions that may experience the impacts of the rapidly changing global environment more strongly than others (Purohit, 1991). Because of their unique characteristics and opportunities, various aspects of global change interactions with mountainous regions have already triggered significant activity amongst the research community. Within the IGBP the programme elements like BAHC and GCTE have developed a number of activities related to mountainous regions (Becker *et al.*, 1994; Chalise & Khanal, 1996). All these developments are related to and confirm the importance of chapter 13 of Agenda 21 (the so-called "Mountain Agenda"), endorsed by the UNCED entitled "Managing Fragile Ecosystems- Sustainable Mountain Development", which also supports the need for further researches (Ives *et al.*, 1997). BAHC and GCTE intend to provide the basic understanding of global change impacts on hydrology and terrestrial ecosystems that will underpin the impact studies undertaken by other groups, many of which will be working within the partner IGBP programme elements such as PAGES, LUCC (IGBP/IHDP) and through the START regional networks around the world.

"Greenhouse gases", especially carbon dioxide, are intimately connected to climate change. The international scientific community has responded to this unprecedented carbon challenge by developing a ten-year Global Carbon Cycle Joint Project. This project is co-sponsored by the International Geosphere-Biosphere Programme (IGBP), the International Human Dimensions Programme on Global Environmental Change (IHDP) and the World Climate Research Programme (WCRP). The three programmes already worked together on some areas of carbon-cycle research. WCRP and IGBP work closely together on climate variability and have established a programme of ocean-carbon measurements. Recently, the IHDP and IGBP jointly sponsored a project on land-use/cover change, including its implications for the carbon cycle. The challenge of obtaining and disseminating global carbon-cycle observations has been taken up by the Integrated Global Observing Strategy Partnership (IGOS-P). The ultimate goal is to understand the system well enough to make reliable projections of carbon-cycle dynamics into the future. In this book, an effort has been made to assess the impact of land-use/cover change on climate change, as it is considered as a major cause with high implications for landscape management.

1.4. THE KYOTO PROTOCOL

The Kyoto Protocol of the United Nations Framework Convention on Climate Change (UNFCCC) is the first step by the world's nations to limit the emissions of carbon dioxide and other greenhouse gases at a level that would prevent human-induced actions from leading to "dangerous interference" with the climate system. In 1997 the Kyoto Protocol was unveiled; dealing predominantly with greenhouse gases, it was the largest and most ambitious piece of environmental legislation ever seen.

The Kyoto Protocol is an international agreement under which developed countries agreed to reduce their overall greenhouse gas emissions to at least 5% below 1990 levels in the commitment period 2008-2012. Article 3 of the Kyoto Protocol provides for making net changes in greenhouse gas emissions by reducing emissions and removing greenhouse gases using carbon sink, in response to direct human-induced land-use changes and forestry activities. The Kyoto Protocol calls for net greenhouse gas emissions and carbon sequestration to be measured as changes in carbon stocks, and also calls for establishing carbon stock baselines. Consequently, the magnitude of changes in above- and below-ground carbon pools needs to be assessed (Johnson & Kern, 2002). Accurate evaluation of the carbon pools and changes in them due to land-use change have been discussed actively since the Kyoto Protocol was formulated (Watson *et al.*, 2000; Lal *et al.*, 2001; Kimble *et al.*, 2002). In the Protocol, 38 of the most developed countries of the world (who collectively emit about 60% of the total carbon emissions) have been given emission reduction targets. These targets have been arrived at by a long negotiation process, which has attempted to take into account the special circumstances of each country. For example, the European Union was given a reduction target of 8%, Japan 7%, and the USA 6%, whilst Australia is allowed to increase its emissions by 8%. The emissions are to be counted in the period 1990 to 2010 (Grace, 2004).

The greenhouse gases that are especially important are: carbon dioxide, methane and nitrous oxide, all of which have been rising fast over the last few decades. Carbon dioxide is the main product of fossil fuel burning, and is quantitatively the most important. There are industrial gases too: hydrofluorocarbons (HFCs), perfluorocarbons (PFCs) and sulphurhexafluoride (SF6), these gases are far from equal in their greenhouse effect, and so to add them up it is necessary to use their Global Warming Potential, as index agreed upon by the IPCC.

To become law, 55 countries must ratify the Protocol, and those 55 must account for 55% of emissions. However, the USA announced its intention not to ratify the Protocol in March 2001. Since USA emits over one-third of the global C-emissions this was a massive set-back for the Protocol. The Protocol allows countries to count the following practices, known as "flexible mechanisms", towards their emissions reductions:

1. Planting new forests and thus creating "sinks" for carbon (from 1990) and adopting new agricultural practices that reduce emissions (Article 3.3 and 3.7).
2. Carbon trading (Article 6 and 17).
3. A Clean Development Mechanism, CDM (Article 12).

1.5. INDIA'S INITIATIVES ON CLIMATE CHALLENGE

In spite of strong traditions of natural preservation and its due recognition in the Indian Constitution (Article 48a), "the State shall endeavour to protect and improve the environment and to safeguard forest and wildlife in the country", the real orientation towards contemporary environmental issues began in the aftermath of the 1972 Stockholm Conference on Human Environment. Between Stockholm Conference and the World Summit on Sustainable Development, India has established a fairly strong organizational structure and legal and policy framework for the protection of environment, focusing on poverty alleviation and natural resource conservation. For India, a formidable issue is the available land/water mass (2.5% of the world) for a rapidly increasing population (16.7% of the world). This creates enormous pressure on natural resources and human development as reflected by poverty and inequality, quality of water resources and environmental health risks (MOEF, 2002 a, b; Singh *et al.,* 2006).

With less than 21% of forest cover, including nearly 42% as degraded forests, we still are far behind the national goal of 33% of total area under forests set by the National Forest Policy, 1988 (Singh *et al.,* 2006). Due to increasing demand of land by ever-growing population, nearly 50% of the country's land is degraded at varying levels. The problems associated with water scarcity are aggravating with every passing day. India is among 17 countries which are most water scarce, and in 2025 will not have enough water to maintain 1990s levels of per capita food production from irrigated agriculture and meet industry, household and environmental needs (Seckler *et al.,* 1999). The wetland systems are now among threatened categories. Similarly is the case of Indian Mountains where the pressures on natural systems are increasing at an alarming rate.

An expert committee on climate change has been formed by the Ministry of Environment and Forests to "study the impact of anthropogenic climate change in India" and "identify the measures that we may have to take in the future in relation to addressing the vulnerability to anthropogenic climate change impact".

Chapter 2

LAND-USE/COVER CHANGE DYNAMICS

2.1. INTRODUCTION

Human actions are altering the terrestrial environment at unprecedented rates, magnitudes, and spatial scales. Land-cover change stemming from human land uses represents a major source and a major element of global environmental change (Turner II *et al.*, 1994). Land is the most important natural resource which embodies soil, water, vegetation and other productive resources upon which all terrestrial bio-systems are dependent. The science of land-use/cover change comprised of identifying and quantifying changes in the landscape, requiring and understanding of that existed (or currently exist) in the landscape, how the landscape will look in near future (prognostic component), and socio-economic forces that drive changes. In the recent time advancement of the technological inputs has made land-cover change monitoring easy. There has been a new earth observation system, satellites with very fine resolution that has helped new opportunities to advance the frontier of land-use/cover change researches. Over the years there has been frequent concern of land resource degradation due to increasing human and bovine population that leads to put high pressure on certain resources and thus changes land-use/cover, leading to loss of soil fertility and depletion of forest resources (Milas, 1984; Toit, 1985; Raphael, 1992). Other factors contributing to resource degradation include the breakdown of traditional resource management systems (Rai *et al.*, 1994; Rai, 2007), inequality in access to natural resources (Baker, 1984; Whitlow, 1988) and commercialization (Singh *et al.*, 1984).

Land-cover change stemming from human land-uses represents a main element of global environmental change. Land-use/cover changes contribute to globally systematic changes because of greenhouse gas accumulation in the troposphere and stratospheric ozone depletion (Turner *et al.*, 1993). Over the past 150 years, for example, land-cover changes and fossil fuel combustion have released approximately equal amounts of carbon dioxide to the atmosphere (Houghton & Skole, 1990). The most spatially and/ or economically important human uses of land globally include cultivation in various forms, livestock grazing, settlement and construction, reserves and protected lands, and timber extraction. These and other land-uses have cumulatively transformed land-cover at a global scale. The consequences have been significant for many aspects of local, regional and global environments, including climate, atmospheric composition, biodiversity, soil condition and water and sediment flows.

Land transformation from forest to other usages are generated due to the quest for fast economic development and expanding agricultural activities has increased the exploitative pressure on the forests in the Himalaya (Singh *et al.*, 1984). A rapid depletion of forest resources has led to environmental degradation and economic deterioration all too widely in the Himalaya. Himalayan forest loss is recognized as a regional and global problem, a little is known about the link between resource use and effects on forest fragmentation and loss. Knowledge of the dynamics and patterns of land-use/cover and forest fragmentation inventories are needed for the long-term sustainability of human-forest interactions and for developing management policies that protect and enhance Himalayan forests.

A better understanding of land-use/cover change is of crucial importance to the study of global climate change because changes in land-cover are caused by land-uses, which, in turn, are governed by human driving forces. Therefore, comprehensive information on the spatial and temporal distribution of land-use/cover categories and the pattern of their change is a prerequisite for understanding the global, region and local climate change. Land, water and vegetation are the most important natural resources of mountain regions and these are much more degraded today than they were in the past.

2.2. DYNAMICS OF GLOBAL LAND-USE/COVER CHANGE

Changes in land-use reflect the history and perhaps, the future of humankind. They are marked with economic development, population growth, technology, and environmental change (Houghton, 1994). Land-use change also occurs in various forms, including both changes in area and changes in the intensity of use. Clearing of forests to yield higher returns from land has a long history. It is estimated that about half of the earth's land area was covered by forests 8000 years ago, as opposed to 30% today (Ball, 2001; FAO, 2001). In recent decades there has been large scale conversion of forests to other land-uses worldwide. Man has emerged as a primary cause of land-cover change around the world. The widespread transformation of land is mainly through efforts to provide food, shelter and products for human use. The effects of land-use change and management are so significant that collectively they form one of the major environmental changes that are occurring at a global scale (Dale *et al.*, 2000). To understand human induced change in land-cover, efforts have been made to quantify the nature and extent of land-use/cover change at a global, national, regional and local scale.

2.2.1. Land-use/Cover Change Detection

The area of land used productively by humans today is approximately 4700 $\times 10^6$ ha (Houghton, 1994; FAO, 1990). Approximately 30% of this land is devoted to crops, including tree crops as well as annual row crops, and approximately 70% is in permanent pasture. These lands together comprise approximately 32% of the land surface of the earth. The principal trends in change in major cover categories are presented in Table 2.1. Over the last three centuries, the total global forest cover and woodlands diminished by 12×10^6 million km^2 (19%), grasslands and pasture by 5.6×10^6 million km^2 (8%); at the same time cropland has increased by 12×10^6 million km^2 (466%) (Table 2.1). Although some of this increase took place in the 19th century as a result of agricultural expansion in the USA, Canada and the former Soviet Union, the most rapid increase occurred in tropical regions after 1950 (Houghton, 1999). Nevertheless, the global expansion of cultivated lands has accelerated with approximately half of the increase occurring in the last 50 years. Such large changes in land-cover can have important consequences such as significant changes in regional and global climate (Dickinson & Henderson, 1988), modification of the global cycles of carbon, nitrogen and water (Houghton *et al.,* 1983) and increased rates of species extinction and biological invasion (Vitousek *et al.,* 1997). The clearing of natural vegetation and cultivation of soils associated with croplands brings a large change to the environment. Upward of 90% of the carbon originally in the vegetation and approximately 25% of that initially held in the soils (to a depth of 1 m) is lost to the atmosphere as a result of this conversion. Less carbon is lost from soils converted to pasture because they are generally not cultivated, but the burning of tropical forests for pastures release carbon monoxide, methane, and nitrous oxide as well as carbon dioxide. Other emissions result from the subsequent use of the lands cleared from forests. Nitrogen fertilizers applied to croplands result in emissions of nitrous oxide, and both paddy rice and cattle ranching increase emissions of methane (Houghton, 1994).

Changes accelerated globally, in terms of both the conversion of lands to cultivation and the intensification of agriculture on land already cultivated. Cropland expansion will undoubtedly continue in the near future, but land-cover modification, through increasing intensification of agriculture, is likely to be of greater importance than further land-cover conversion (Ruttan, 1993). Irrigation of cropland has expanded some 24-fold over the past 300 years, with most of that increase taking place in the 20th century. This practice has increased methane emissions, while the increasing frequency of land tillage worldwide has affected soil carbon (Cole *et al.*, 1989).

Table 2.1. Global human-induced conversions in selected land-use/covers from 1700-1980

Land-use/cover	Year	Area ($\times 10^6$ km^2)	Year	Area ($\times 10^6$ km^2)	% change
Cropland	1700	2.65	1980	15.01	+466
	1700	3.00	1980	14.75	+392
Irrigated cropland	1800	0.08	1989	2.00	+2400
Closed forest	pre-agricultural	46.28	1983	39.27	-15.1
Forest and woodland	pre-agricultural	61.51	1983	52.37	-14.9
Grassland/pasture	1700	68.60	1980	67.88	-1.0
Lands drained			1985	1.606	
Urban settlement			1985	2.47	
Rural settlement			1990	2.09	

Source: Richards 1990.

2.2.1.1. Changes over the Last Century

Beginning of the eighteenth and nineteenth centuries, a different approach for determining changes in land-use is possible. Data can be obtained directly from the land-use statistics. Croplands are relatively well documented in census records worldwide (Richards, 1990). Based on the historical data and assumptions, approximately 28% of the forest area in Latin America was lost between 1850 and 1985 (Houghton *et al.*, 1991). The area in croplands, pastures, and fallows had grown from 357×10^6 ha to 918×10^6 ha over the 135-year period. A similar study in South and South-east Asia showed a 34-38% reduction in forest area over the last 140 years (Houghton & Hackler, 1994; Richards & Flint, 1994). Lands altered for human use in that area expanded by approximately 176×10^6 ha. The loss of forests and the expansion of agricultural area occurred in a similar pattern in Tropical Africa over the last century (Houghton, 1994). Before 1960, croplands were expanding more rapidly in regions outside the tropics. In North America, Europe, the former Soviet Union, and China, the largest changes in land-use occurred earlier (Houghton & Skole, 1990; Williams, 1990), sometimes much earlier (Darby, 1956).

In Indian context, particularly Central Himalaya, the massive extension of agricultural activities appears to have begun in 1823 when the British decided to expand the amount of arable land, and by the late 1860s in Kumaun the cultivated land had more than doubled. The 1840s and 1850s constituted the first era of large-scale uncontrolled deforestation to meet the timber demands of the expanding cities on the Indo-Gangetic plains (Singh *et al.*, 1984). The commercial exploitation of forests has since continued along with the expansion of agriculture.

2.2.1.2. Changes over the Last Decade

In the last three decades, with launch of Remote Sensing Satellites, direct measurement of areas of different types of land-cover and of changes in this cover has become possible. Overall, estimates for closed forests seemed to agree fairly well for the entire tropics (approximately 7.5×10^6 ha yr^{-1}). The FAO/UNEP survey (1981) reported the area of fallow to 50% of the deforestation of closed forests in Tropical America, Africa and Asia, respectively, to shifting cultivation. Myers (1980; 1991), on the other hand, believed that shifting cultivation was largely being replaced by permanently cleared land and that the area of fallow was decreasing. He estimated that approximately 10×10^6 ha of fallow forests were cleared for permanent use. According to Myers (1991), the annual loss of closed forests has almost doubled from 7.34×10^6 ha in 1979 to 13.86×10^6 ha in 1989. FAO's recent estimate (1993) is also higher (approximately a 50% increase) (Table 2.2).

According to these tropics-wide studies of deforestation, the major agents of deforestation are farmers clearing land for either shifting or permanent cultivation, sometimes settling in forests made accessible through logging. Thus, the main cause of deforestation could be considered to be agriculture. One other change in land-use, increasingly important in recent decades, is the growth of urban areas. Although the fraction of the earth covered by urban areas is less than 1% of the land surface, the sprawl of suburban areas is displacing both agricultural and natural ecosystems. Furthermore, emissions of greenhouse gases from transportation are large and concentrated in urban and suburban areas.

Table 2.2. Estimates of rates of tropical deforestation (10^6 ha^{-1} yr^{-1})

Estimate	America	Africa	Asia	All tropics
Independent estimate for the late 1970's[*]				
FAO/UNEP (1981)	4.12	1.33	1.82	7.27
FAO estimates[**]				
1976-1980	5.61	3.68	2.02	11.30
1981-1990	7.40	4.10	3.90	15.40
Percent Increase	32	12	93	36
Myers estimates[$]				
1979	3.71	1.31	2.58	7.60
1989	7.68	1.58	4.60	13.86
Percent Increase	107	21	78	82
1989 (revised)[#]	4.48			10.66
Percent Increase (revised)[#]	21			40

[*] Closed forests only.
[**] From FAO/UNEP 1981, FAO 1993; Closed and open forests.
[$] Myers 1980, 1991; closed forests only.
[#] Revised rates and percent increase are based on a rate of deforestation in Brazil of 1.8 × 10^6 rather than 5.0× 10^6 ha^{-1} yr^{-1}.
Source: Houghton 1994.

2.2.2. Extent of Land-use/Cover Change in Indian Parts of Himalaya

Himalaya constitutes a unique geographical and geological entity comprising a diverse social, cultural and environmental set up. Geological instability interacting with a complex of problems including population pressure, deforestation, landslides, erosion, out migration and poverty manifest fragility to the Himalayan ecosystem. Alteration in the environment and patterns of development in the Himalaya in future are considered crucial. Himalaya influences even the climatic changes in the continent. Himalaya controls monsoonic climate and ensures adequate water flow in major rivers in a part of South Asia.

More than 2800 km in length and 220 to 300 km wide, the Indian Himalayan region is spread over the Jammu & Kashmir to Arunachal Pradesh. It has a total geographical area of approximately 591000 km^2 (18% of India) inhabited by 51 million persons (6% of India) (Anonymous, 1992). Agriculture is the primary occupation of the people all through the region but agricultural land-use patterns vary from region to region. While in the north-eastern region, shifting cultivation continues to be the general practice. Indian Himalayan Region (IHR), forest cover is the major land-use, which covers about 52% of the total reporting area of the region followed by wastelands. The forest cover in the region has recorded marginal increase (about 0.41%) during the period 1999-2001 where as the country's forest cover has recorded a significant growth (about 6%) during the same period in spite of rapid urbanization. This could be because of high dependency on forest in the IHR than the arable land that contributes only about 11% of its total reporting area.

The people of the Himalayan region, especially those in the north-east, over the years had adopted traditional practices of replenishment in the region. One such method is *Jhum* (slash and burn) cultivation, which is basically "rotational bush fallow" agriculture. This traditional tribal practice enabled regeneration of forests before the same land was cultivated again. The *Jhum* cycle was once considered to be as long as 30 years, but in the recent past, studies have shown that the cycle has shrunk to as short as 2-3 years. As the *Jhum* cycle becomes successively shorter, the rate of soil erosion gets accelerated. This is a strong indicator of the deteriorating ecological balance of the region and is also a statement on the increasing human pressure on land and growing food needs. Table 2.3 revealed the major land-use pattern in the IHR.

The north-eastern region of India comprising eight states viz, Assam, Arunachal Pradesh, Meghalaya, Manipur, Mizoram, Nagaland, Tripura and Sikkim, has a total geographical area of 262180 km^2, which is nearly 8% of the total area of the country with more than 39 million population. Nearly 70% of the total area is hilly and shifting cultivation is the chief land-use practice in the hilly regions. The region is characterized by different terrain, wide variations in slope and altitude, land tenure systems and cultivation practices.

The region is well known for its forest wealth. Five major forest types in northeast India were identified viz., Tropical forests, occurring up to elevation below 1000m, contain several economically important trees, including *Dipterocarpus macrocarpus, Mesua ferrea, Shorea assamica* and *Terminalia myriocarpa*. Subtropical forests, located between 1000-1800 m, are dominated by species of *Castanopsis, Ficus* and *Quercus*. Pine forests, extending into both the subtropical and temperate belts are represented by several *Pinus* species. Temperate forests, occurring as a continuous belt between 1800 and 3500 m elevation are broad-leaved at the lower end and coniferous at the upper end of their elevation range. Alpine forests, which occur on peaks above 4000m, contain several Rhododendron species. Tropical semi-evergreen forests occur along the foothills and river banks up to 600m throughout the region. Degraded forests, bamboo forests and grasslands are a result of anthropogenic factors such as shifting cultivation and excessive grazing and of natural factors such as landslides and fire.

Table 2.3. State wise major land-use distribution pattern in the Indian Himalayan Region

States	Geographical Area (km²)	Per cent of Area		
		Agricultural land	Wasteland	Forest land
Jammu & Kashmir	222236*	4.7	64.6	9.6
Himachal Pradesh	55673	14.5	56.9	25.8
Uttarakhand	53483	12.5	30.1	44.8
Sikkim	7096	16.1	50.3	45.0
W. Bengal hills	3149	43.5	2.2	69.7
Meghalaya	22429	48.2	44.2	69.5
Assam hills	15322	10.5	56.6	79.8
Tripura	10486	29.6	12.2	67.4
Mizoram	21081	21.2	19.3	83.0
Manipur	22327	7.3	58.0	75.8
Nagaland	16579	38.4	50.7	80.5
Arunachal Pradesh	83743	3.5	21.9	81.3
India	3287263	55.8	20.2	20.6

* Included 78,114 and 37,555 km² occupied by Pakistan and China, and 5180 km² handed over by Pakistan to China.

Source: Wastelands Atlas of India, 2000 & FSI 2000.

Administrative control regimes for forest cover in northeast India is unlike that in other regions of the country (Table 2.4). The existing protected area network in north-eastern region covers approximately 15579.10 km² and contains 4 biosphere reserve, 13 national parks and 43 sanctuaries. Also, reserve forests constitute approximately 25% of the region. Biosphere reserves, national parks and sanctuaries are legally protected in India and offer the highest levels of protection to biodiversity. Reserve forests are managed by the state forest department for a variety of purposes, including selective logging for timber harvesting. Most of the forested area in North-East region (29.88% of total forests) is "unclassified state forests". Pant (1997) describes the legal status and use of unclassed state forests and the role of traditional institutions such as village councils in managing such forests. Unclassed state forests do not have a clearly defined legal status; community control of these forests is implied and accepted by the state and local people.

Table 2.4. Administrative classifications of forest cover in North-East India (km^2), (Poffenberger, 2005)

States	Geo-graphic area	Reserved	Protected	Un-classed	Total	Per cent to total geographical area
Arunachal Pradesh	83743	15300	4200	32000	51500	61.50
Assam	78438	18100	2114	8900	29114	37.12
Manipur	22327	1400	4100	11800	17300	77.48
Meghalaya	22429	700	300	8500	9500	42.35
Mizoram	21081	7100	3600	5200	15900	75.42
Nagaland	16579	300	500	7800	8600	51.87
Sikkim	7096	2261	265.10	104	2630.1	37.06
Tripura	10486	3600	500	2900	7000	66.75
North-Eastern Region	262179	48761	15579.10	77204	141544.10	53.98

2.2.2.1. Land-use/Cover Change in Sikkim Himalaya: A Watershed Case Study

Sikkim a small hilly state of India in the eastern Himalayan biogeographic zone that harbors largest number of endemics and endangered species in the Indian subcontinent, is recognized as one of the biodiversity 'Hot Spot' of global significance (Khoshoo, 1992). The entire Sikkim Himalaya constitutes a mountainous terrain spreading over 7096 km^2 and is quite well-known in terms of its resplendent flora and faunal aggregation. The entire state is divided into two catchments, i.e., the Tista and the Rangit catchment. Further, the Rangit catchment comprises 51 micro-watersheds, including the Mamlay watershed, which has been selected as the case study.

The land-use/cover pattern of the Mamlay watershed varies considerably depending upon the ecological conditions, altitude, lithology and slope aspect. Apart from these factors, population and livestock growth, urban sprawl also affected the land-use/cover pattern. The land-use/cover data generated through satellite imageries for the year 1988 and 2001 has been classified into four major classes i.e., (i) forests (ii) agroforestry (iii) agricultural land, and (iv) wasteland.

Forest constitutes about 42% of the total area of the Sikkim Himalaya. Records show that the area under closed canopy was only about 14% (Sudhakar *et al.*, 1998). The forests of the State have suffered a serious setback during recent years due to tremendous pressure arising out of ever increasing demand for fuel wood, fodder and timber coupled with diversion of forest lands to non-forest uses in the name of developmental processes. Forest is the most important land-cover on higher steep slopes and ridges. The forest land includes temperate natural forest dense, temperate natural forest open and sub-tropical natural forest open. The total forest land in the Mamlay watershed accounts for 69% and 49% of the total areas of the watershed in 1988 and 2001, respectively (Table 2.4) (Sharma, 2003). The spatial distribution pattern shows that the northern, western and eastern parts of the watershed area are dominated by dense mixed and open mixed forests. Some forest blanks are also found in the reserved forest categories, this indicates the high human and livestock pressure in the area.

The agroforestry practices in the watershed are traditional and promising for higher economic returns. Two types of agroforestry systems are very common in the watershed, i.e., (i) large cardamom based, and (ii) mandarin orange based. About 4% areas came under agroforestry practices in 1988 and 2001, respectively (Table 2.4). Large cardamom (*Amomum subulatum* Roxb.) is the most important perennial cash crop in the Sikkim Himalaya. Cardamom is predominantly farmed in between 600-2000m elevations. The plant is a shrub by habit and has several tillers consisting of pseudostems with leaves on the upper part. Cardamom is a shade loving crop grown under forest cover. Himalayan alder (*Alnus nepalensis*, D. Don) is the most common species used as shade tree in cardamom farming.

Agricultural land includes built-up land, rainfed and irrigated land. About 14.39% and 30.53% area was under this category in 1988 and 2001, respectively. The land-use/cover pattern in the watershed as a whole showed about 2% area under built-up land in both the years. This includes only cluster settlements. The low lands along the river bed, commonly known as khet, are irrigated and paddy cultivation is the common practice in this land. The whole watershed had only about 2% of its area under irrigation in both the assessment years. Rainfed known as pakho cultivation covered about 12 and 28% area in 1988 and 2001, respectively. This land is suitable for cultivation of mainly maize, ginger and pulses (Sharma, 2003).

Four types of crop rotations are found in the watershed viz., maize-pulse/maize-potato/maize-ginger-pulse, these are practiced in rainfed conditions and one type (maize-paddy-fallow) in irrigated areas of the watershed (Rai, 1995).

The maize-pulse crop combination is quite common and maize is harvested much earlier than pulses. Potatoes are becoming more popular as a cash crop. They are usually grown in triple cropping rotations after monsoon maize and relay-cropped with pulses or ginger. The intensity of cropping varies from farm to farm and from household to household due to differences in socio-economic conditions, particularly inputs and products, dependence on land and tenurial system etc.

Wastelands category includes rock outcrops, landslides, forest blanks, degraded forests and scrublands. The wasteland covered about 11 and 15% of the total area of the watershed in 1988 and 2001, respectively (Table 2.4). In 2001 about 9 ha area was under landslides whereas no landslides were observed in 1988. The surface water bodies include river, streams and springs. Owing to vegetation cover over the major drainage channels of the watershed, satellite imagery does not show a clear response for these channels in terms of the spectral signature of the water bodies.

The land-use/cover change detection was generated by the multi-date satellite data. Monitoring of land-use/cover reflected that changes were greater in extent over the span of 13 years in the land under different categories. Table 2.4 is a summary of changes in land-use between 1988 and 2001. The most dramatic changes are the increase in agricultural area and decrease in forest cover area. The open cropped area sub-tropical increased by more than 166% for the thirteen years period, while wasteland subtropical increased by about 117%. The total forests cover comprising temperate dense mixed, open mixed, and sub-tropical open mixed forest decreased by 28% during 1988-2001 (Table 2.4). This reflect the conversion of dense mixed forest to open mixed forest to degraded forest and dense mixed forest with agroforestry to open mixed forest with agroforestry and further to open cropped area. Ground-truth verification supports the finding that the depletion of closed forest or its conversion into other categories is the result of maximum anthropogenic pressure on the limited forest resources (Sharma, 2003).

Population growth is important because the major change in land-use, globally, as well as in Hindu-Kush Himalayan region, has been the expansion of agricultural land, at least in part for food at the cost of forest. Field checks, interviews and detailed households survey in the watershed revealed that a large number of woody species are utilized for fuel, fodder and timber for house construction and in making agricultural implements. Most of these species are collected from the forest and agroforestry systems. Each household on an average is composed of 6 persons and consists of 4 cattle. Fuel wood consumption per house hold is as much as 21 kg per day, through a minimum of 4000 kg (range 4000-5800kg) of dry firewood annually (Table 2.5). Annual consumption of firewood was greater for cooking (69%) followed by animal food preparation

Suresh C. Rai

(9%), water heating (7%), house warming (6.7%), local wine /beer preparation (6%) and use for festivals (2%) (Sundriyal & Sharma, 1996).

Table 2.5. Area under different land-use/cover and change detection of the Mamlay watershed based on remote sensing data, 1988-2001(based on Sharma & Rai, 2007)

Land-use/cover	Year				Variation	
	1988		2001		(1988 - 2001)	
	(ha)	(%)	(ha)	(%)	(ha)	(%)
Forest						
Temperate natural forest dense	606.81	20.13	160.00	5.32	- 446.81	-73.63
Temperate natural forest open	815.09	27.04	982.24	32.59	167.15	20.51
Sub-tropical natural forest open	681.54	22.61	362.25	12.03	- 319.29	-46.85
Agroforestry						
Cardamom based agroforestry	114.78	3.81	114.78	3.81	--	--
Mandarin based agroforestry	17.42	0.58	17.42	0.58	--	--
Agriculture*						
Open cropped area temperate	243.89	8.09	413.62	13.72	169.73	69.59
Open cropped area sub-tropical	189.96	6.30	506.33	16.81	316.37	166.55
Wasteland**						
Wasteland area temperate	341.99	11.35	451.92	14.99	109.93	32.14
Wasteland area sub-tropical	2.5	0.08	5.42	0.18	2.92	116.80
Total	3014	100	3014	100		

* Irrigated area, rainfed area and built-up area.
** Degraded forest, scrub land, forest blanks, rock out crops, land slides.
-- denotes no change.

Each family maintains four animals consisting of cattle, pig and goat. Fodder is collected mainly from the forest (65%) and from agricultural fields (35%). Average fodder collection per household varies from 5000-6500 kg yr^{-1} (mean 5700 kg per household yr^{-1}) from the forest area (Table 2.5). Most of the tree sprouts as well as ground herbaceous vegetation are removed for fodder purposes. In addition, unpalatable species and leaf litter are used for animal bedding.

Table 2.6. Per household consumption of fuel wood, fodder and timber in the Mamlay watershed (based on Sundriyal &Sharma, 1996)

Fuel wood consumption	
Daily requirement per household	15-21 kg
Annual requirement per household[*]	4000-5800 kg
Fodder collection	
From forest[**]	5700 kg household^{-1} year^{-1}
From agricultural fields[**]	2630 kg ha^{-1}
Timber (on per household basis)	
Small size poles (bamboo)	
Purpose	cattleshed, baskets, mats, minor repairs
No. of poles required	100+
Average size of poles (CBH[***])	<30 CM
Time interval of need	5 - 7 years
Medium size pole	
Purpose	house repairs, furniture etc.
Wood volume required	2.5 – 4.2 m^3
Time interval of need	15 – 20 years
No. of trees required	20-40
Average size (CBH)	30-90 cm
Large size poles	
Purpose	new house construction
Wood volume required	5 - 7 m^3
No. of trees required	7 – 10
Average size (CBH)	90 – 125 cm
Time interval of need	20 – 30 years

[*] large cardamom growers use an additional 70 to 80 kg of fuel wood (dry weight) for curing per 100 kg of cardamom.
[**] on dry weight basis (fresh weight-dry weight ratio is 3:1).
[***] CBH is pole's circumference at breast height.

Family fragmentation every 20-25 years leads mostly to construction of many new houses and almost all houses are made of wood. Generally a space of two rooms needs 3-6 m^3 wood, depending upon the socio-economic condition of the farmers, and thus a huge amount of wood is collected each year. Field observations revealed that a tree of 50-90 cm and 90-125 cm CBH produces about

0.3-0.4 and 0.8-1.0 m^3 wood respectively. Generally large timber poles are harvested for making ceilings, doors, windows and beams. Medium size poles are used in making furniture and repairs, whereas small size poles for making cattle sheds or temporary huts (Table 2.5).

Human and livestock population pressure on the limited land resources has increased in recent years. This has resulted from the construction of road, fuel, fodder and timber extraction, encroachment into forests and more land utilization for agricultural expansion. Increased pressure on forests has brought tremendous changes in the pattern of land-use including reduction in forest cover. The expansion of agricultural land can be mainly attributed to fragmentation of upland farm families, which has led to expansion of the agricultural area. A bulk of settled agriculture fields in the watershed occurs on sloping terraces along the steep hill sides. Slopes of some agricultural land exceed 40° but most of them fall in between 20 to 35° (Rai *et al.*, 1994). The overall pattern of the forest and agroforestry revealed that all sites are under increasing biotic pressure from the neighboring villages. Now there are trends/evidences of indiscriminate cutting and mismanagement during recent years which has resulted in more damage of some species than others.

The forest and agroforestry systems has been meeting and satisfying the material needs of the majority of the population of the watershed, but now evidence of decline in species number and composition are emerging and it is apparent that local subsistence needs are causing much of the degeneration in the forests. Indiscriminate cutting by people, selective felling by Forest Department, plantation of exotic species like *Cryptomeria japonica*, and use of enormous amounts of wood in house construction and large scale cardamom curing are the most common causes of forest destruction. Field visits revealed that the cutting process is highly irregular in both space and time and a huge amount of wood is wasted. Nearly 40-50% and 20-30% is wasted in the processes of timber and fuel collection, respectively. Generally, smaller dry and dead, and fallen logs and branches are not collected and trees of smaller girth classes are preferred due to easy extraction.

Interviews with the residents of the watershed revealed that previously the forest had a good number of individuals of *Michelia excelsa, M. lanuginosa, Juglans regia, Cedrela toona*, and various other timber trees. The most significant extraction of these species was done after 1970 due to construction of more luxurious house throughout the State. In Sikkim as well as in the Mamlay watershed it has been observed that the rate of immigration into the area was high during 1971-1991. Most of the immigrants were traditionally cultivators (personal observation). The growth of population in the watershed was at an average rate of

2.84% per year over the period of 1981-1991 and the forest cover decreased at an alarming rate of upto 2.80% per annum, which is excessively high. Thus the family fragmentation and population pressure is continually taking place within the watershed, which indicates that the man-land ratio is likely to further decline in the near future. Considering all these factors, it can be said that this natural forest stand is at severe risk of reduction which may lead to the disappearance of many species in the near future.

Lack of effective land-use planning and uncontrolled population growth has contributed to the present deplorable state of affairs. In general, the area shows increasing environmental degradation and resource depletion, while very little conservation efforts are being made to reverse the trend. These results indicate that a sustainable land-use/cover management plan is urgently needed for the area.

Chapter 3

GLOBAL BIOGEOCHEMICAL CYCLES

3.1. INTRODUCTION

Atoms and molecules move through ecosystems under the influence of both physical and biological processes. The pathways of a particular type of matter through the Earth's ecosystem comprise a "Biogeochemical cycle" (Strahler & Strahler, 2003). In each biogeochemical cycle, any area or location of concentration of a material is a *Pool*. There are two types of pools: *active pools*, where materials are in forms and places easily accessible to life processes, and *storage pools*, where materials are more or less inaccessible to life. A system of pathways of material flows connects the various active and storage pools within the cycle. Pathways can involve the movement of material in all three states of matter- gas, liquid, and solid. For example, carbon moves freely in the atmosphere as carbon dioxide gas and freely in water as dissolved CO_2 and as carbonate ion. It also takes the form of a solid in deposits of limestone and dolomite (calcium and magnesium carbonate).

Of the 90-odd chemical elements occurring in nature, 30 to 40 are known to be required by living organisms. Out of these, six most important chemical elements necessary for life are: carbon, hydrogen, oxygen, nitrogen sulphur and phosphorus. The first three are combined in energy-rich materials such as carbohydrates and fats. These together with nitrogen and sulphur are essential ingredients in all proteins. Phosphorus is needed for the transfer of chemical energy within protoplasm where this energy is used for activity or growth (Singh *et al.*, 2006). The biogeochemical cycles are driven by energy flow and are crucial for the maintenance of life on earth in its present form, and biological processes largely determine the main features of these cycles.

Terrestrial ecosystems are important components in the biogeochemical cycles that create many of the sources and sinks of carbon dioxide, methane, and nitrous oxide and thereby influence global responses to human-induced emissions of greenhouse gases. The dynamics of terrestrial ecosystems depend on interactions between a variety of biogeochemical cycles, particularly the carbon cycle, the nutrient cycles, and the circulation of water all of which directly affect the climate.

3.2. THE GLOBAL CARBON CYCLE

The emergence of life on earth has led to the conversion of carbon dioxide in the atmosphere and carbon dissolved in the oceans into innumerable inorganic and organic compounds on land and in the sea. The development of different ecosystems over millions of years has established patterns of carbon flows through the global environmental system. Natural exchanges of carbon between the atmosphere, the oceans, and terrestrial ecosystems are now being modified by human activities, primarily as a result of fossil fuel burning and changing land-use.

Human intervention in the global carbon cycle has been occurring for thousands of years. However, only over the last two centuries have anthropogenic carbon fluxes become comparable in magnitude with the major natural fluxes in the global carbon cycle, and only in the last years of the 20th century have humans widely recognized the threat of adverse consequences and begun to respond collectively (GCP, 2003). The carbon cycle is probably the most important for two reasons. First, all life is composed of carbon compounds of one form or another. Second, human activities are modifying the carbon cycle in important ways. The carbon cycle has received particular attention because 60% of the observed global warming is attributable to the increase in carbon dioxide concentration from about 280 μmol mol^{-1} in the pre-industrial period to today's 380 μmol mol^{-1}. In fact, the realization of a rising trend in the carbon dioxide concentration of the atmosphere seems to have first been made as long ago as 1896 by Swedish chemist Arrhenius. He measured the concentrations of CO_2 in ocean and atmosphere, and noting that the ocean had a slightly lower concentration than the atmosphere he inferred the presence of an ocean sink (Grace, 2004). Modern recording of the atmospheric CO_2 signal was started in 1958 by Charles Keeling of the Scripps Oceanographic Institute in the USA. These data are of paramount importance. They were first used to demonstrate that only about half of all the CO_2 emitted from fossil fuel burning remains in atmosphere, and by inference, that there must be carbon "sinks" in the ocean or on land. It was later observed that this airborne fraction varies from year to year, and the international variability is associated with variations in the climate, particularly those caused by EL Nino and major volcanic eruptions (Grace, 2004).

Accounting for about 49% of organism's dry weight, carbon is only next to water in significance to the living world. The major carbon reservoirs and fluxes are given in Table 3.1.The distribution of carbon in major components of the biosphere, as estimated for 1995 is shown in Figure 3.1. The carbon cycle is characterized by small but active atmospheric pool (about 750 Gt; 1Gt, gigaton= 1000 million tons= 1 petagram, Pg or 1×10^{15} g). The cycle involves a gaseous phase, the atmospheric CO_2. The basic movement of CO_2 is from the atmospheric reservoir to plants (by photosynthesis or chemosynthetic fixation) and animals (directly from atmosphere by biochemical fixation, and indirectly by herbivory), and from them to decomposers and then back to atmosphere. There is a reservoir of carbon in the ocean, in addition to atmosphere. The oceans contain more than 50 times as much carbon (40×10^3 Gt) as the air (Figure 3.1) and act as regulator of carbon in the atmosphere.

Table 3.1. Major carbon reservoirs and fluxes, 2004

Reservoirs	Pg Carbon
Atmosphere	800
Biomass	˜ 500
Soil	˜ 1500
Coal	5000 to 8000
Oil	˜ 270
Natural Gas	˜ 260
Unconventional Fossil Fuel	15000 to 40000
Fluxes	Tg Carbon yr-1
Fossil fuel	7*
Natural Gas	13.7
Oil	20.2
Coal	25.5

* Pg C yr^{-1}
Source: ICDC, 2005.

Plants are the major force behind the global carbon cycle, as they fix carbon through the process of photosynthesis. The carbon fixed by plants is used for their own growth and then enters the food chains as animals eat the plants. About twice as much carbon dioxide is taken up by photosynthesis on land as in the oceans and rivers. Terrestrial vegetations are estimated to obtain about 250 times as much carbon as aquatic plants, most of it in the form of forest growth. In the oceans, only a small proportion of the carbon is held in living organisms. Most of it is in the detritus left behind by these organisms, as well as some dissolved carbon dioxide mainly in the form of bicarbonates.

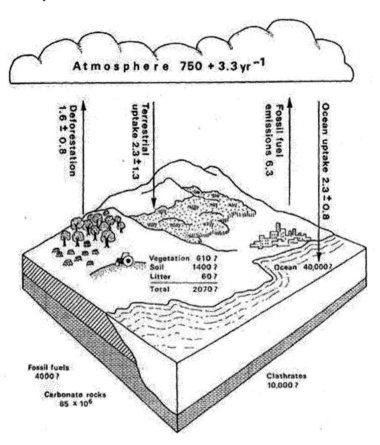

Figure 3.1. Carbon sources and sinks. Approximate carbon stocks are shown in units of Gt of carbon; net fluxes are shown by arrows in Gt of carbon per year. Net photosynthesis of the land surface is believed to be made up of about 120 Gt yr^{-1} of gross photosynthesis and about 60 Gt yr^{-1} of autotrophic respiration, whilst heterotrophic respiration (decomposition) is about 60 Gt yr^{-1}. The sinks shown in the diagram are consistent with inferences from atmospheric concentrations (Grace, 2004).

Respiratory activity of plants and animals, and more importantly, of the decomposers in their processing of dead remains of other organisms and the waste materials result in the return of biologically fixed carbon as CO_2 to the atmosphere. At the global level, the total respiration and decomposition are thought to be equal to total photosynthesis. The fire, by burning the organic materials, provides (non-biological combustion) another route of carbon return from the biota to the tune of about 2-4 Gt per year. The characteristic turn over times vary from a few weeks for marine biota to about 1000 yrs for the CO_2 in the ocean (Singh *et al.*, 2006).

The geological components involving deposition of organisms as coal, oil and animal remains make up a significant constituent of the system. The geochemical part of the carbon cycle stores carbon for a far longer period of time and comprises two main pools: the deep ocean water, with an estimated 38000-40000 billion tones, and rocks, especially limestone, which hold an estimated 66-100 million billion tones (Figure 3.1). Other rocks also lock away carbon, notably those that produce fossil fuels. This carbon too originates in living things-plants provide the carbon for coal, whereas oil and gas also come from marine animals.

The lithosphere including continental and oceanic crusts (carbonate and non-carbonate sediments, and igneous and basaltic rocks) account for about 65×10^6 Gt carbon (Grace, 2004; Singh *et al.*, 2006). On weathering and dissolution of carbonate rocks, the combustion of peat, coal and oil (fossil fuels), and volcanic activity involving deposits of carbonate rocks and fossil fuels, the bound carbon is returned to atmospheric-aquatic reservoir of carbon cycle. The characteristic turn over time for C held in the lithosphere is hundreds of millions of years (Singh *et al.*, 2006).

3.2.1. Human Impact on the Carbon Cycle

One-third to one half of the earth's land surface has been transformed by human action. Many continuing transformations exchange woody and herbaceous plants, including deforestations, desertification, and woody plant invasion (the expansion of woody species into grasslands and savannas). Shifting dominance among herbaceous and woody vegetation alters primary production, plant allocation, rooting depth and soil faunal communities, potentially meters underground, in turn affecting nutrient cycling and carbon storage (Jackson *et al.*, 2002).

The most important human impact on the carbon cycle is the burning of fossil fuels. Another important human impact lies in changing the earth's land covers.

The conversion of forest to agricultural land has varying influence on terrestrial carbon inventories, depending on the type of land-use/covers undergoing change and the post conversion land management. The forest-dominated areas are consequently converted into cultivated land which leads to a considerable loss of soil organic matter and microbial biomass (Hass *et al.*, 1957; Adams & Laughlin, 1981; Bauer & Black, 1981).

Woodwell *et al.,* (1978) claimed that as a result of human disturbance, the terrestrial ecosystems were a net source of carbon to the atmosphere. Simply put, of the 6.3 Gt C emitted annually by burning fossil fuels and removing forest, only about 3 Gt C appeared in the atmosphere. Of the remaining 3.3 Gt C some 2 Gt C was dissolved annually in the ocean, still leaving 1.3 Gt C unaccounted for, and presumed to be absorbed by the terrestrial ecosystems of the world. The important question was, where? In the early 1990s the search for the "missing sink" gathered pace.

If the value of 3 Gt yr^{-1} in carbon uptake is correct, the amount of terrestrial biomass must be increasing at the rate. However, forests are presently diminishing in area as they are logged or converted to agricultural land or grazing land. This conversion is primarily occurring in tropical and equatorial regions, and it is estimated to release about 1.6± 0.8 Gt yr^{-1} of carbon to the atmosphere. Since this release is included in the net land ecosystem uptake of 0.7 Gt yr^{-1}, the remainder of the world's forests must be taking up the 1.6 Gt yr^{-1} losses from deforestation as well as an additional 0.7 Gt yr^{-1}. So we can estimate that mid and high latitude forests are increasing in area or biomass to fix carbon at a rate of 1.6+0.7=2.3 Gt yr^{-1} (Grace, 2004).

The problem of net CO_2 release as a consequence of forest clearing and expanding agriculture perhaps now largely lies in the tropics. It may be pointed out that forests are important carbon 'sinks'. On the other hand, in agriculture a greater release of CO_2 into atmosphere than the removal in photosynthetic fixation especially results from frequent ploughing and from the fact that plants are active only for a short period within an annual cycle. The release of CO_2 as a consequence of rapid oxidation of humus has also other effects. Because of rapid water runoff in landscapes dominated by agricultural fields (instead of water filtering down through humus layers in soil, which normally is the case in forests), the rate of dissolution of many trace elements has declined. Consequently, the need for adding trace elements in agroecosystems in order to maintain the yield is being increasingly felt (Singh *et al.*, 2006).

3.3. THE NITROGEN CYCLE

The nitrogen cycle is the biogeochemical cycle that describes the transformations of nitrogen and nitrogen-containing compounds in nature. It is a gaseous cycle. Nitrogen is the major component of earth's atmosphere. It enters the food chain by means of nitrogen-fixing bacteria and algae in the soil (Figure 3.2). This nitrogen which has been fixed is now available for plants to absorb. These types of bacteria form a symbiotic relationship with legumes, these types of plants are very useful because the nitrogen fixation enriches the soil and acts as a natural fertilizer. The nitrogen fixing bacteria form nitrates out of atmospheric nitrogen which can be taken up and dissolved in soil water by roots of plants. Then, the nitrates are incorporated by the plants to form proteins, which can then be spread through the food chain. When organisms excrete wastes, nitrogen is released into the environment.

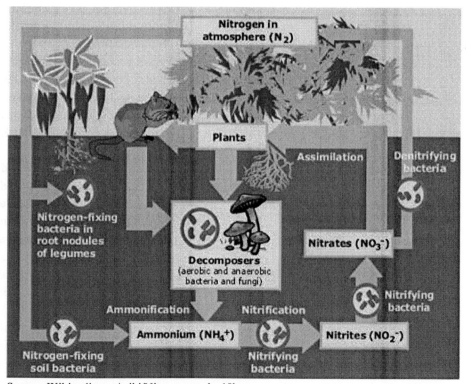

Source: Wikipedia.org/wiki/Nitrogen-cycle-45k.

Figure 3.2. Schematic representation of the flow of nitrogen through the environment.

Nitrogen is an essential component of proteins, genetic material, chlorophyll and other key organic molecules. All organisms require nitrogen in order to live. It ranks fourth behind oxygen, carbon and hydrogen as the most common chemical element in living tissues. Until human activities began to alter the natural cycle (Figure 3.2), however, nitrogen was only scantily available too much of the biological world. As a result, nitrogen served as one of the major limiting factors that control the dynamics, biodiversity, and functioning of many ecosystems. The significance of nitrogen lies in its being the major builder of proteins and nucleic acids, the essential constituents of organisms and regulators of biological functions. Although we live in an atmosphere which is largely N_2 (about 79%, total of 39×10^5 Pg), the gaseous N_2, unlike CO_2, cannot be directly used by most organisms. Atmospheric nitrogen enters the protoplasm of plants only when it is converted into usable forms, viz., nitrate (NO_3) and ammonia (NH_3).

Nitrogen-fixing organisms include a relatively small number of algae and bacteria. Many of them live free in the soil, but the most important ones are bacteria that form close symbiotic relationships with higher plants. Symbiotic nitrogen fixing bacteria such as the Rhizobia, for instance, live and work in nodules on the roots of peas, beans and other legumes. These bacteria manufacture an enzyme that enables them to convert gaseous nitrogen directly into plant-usable forms.

The biological N_2 fixation on global basis is about 44-200Tg (1 Tg= 1×10^{12} g) on land and 1-130 Tg in ocean, atmospheric fixation 0.5-30Tg and industrial fixation 60Tg yr^{-1}. A group of bacteria called nitrite- and nitrate-bacteria convert NH_3 into NO_3, in a process referred to as Nitrification. The NO_3-N is absorbed by the roots of higher plants and is subsequently bound in the protoplasm of organisms as protein and other compounds. From plants, the nitrogen in the bound form moves to animals, and both from plants and animals to detritus.

Several bacteria and fungi participate in the break-down of protoplasm. As decomposition proceeds, amino acids and organic residues are from these NH_3 is released in the process of ammonification performed by a group of bacteria called ammonifying bacteria. This NH_3 is either directly taken up by higher plants, or is converted into NO_3 by nitrification and then taken up by plants or is denitrified.

One can guess easily that if the cycle of nitrogen were to be complete then the NO_3 must return to the atmosphere. The denitrifying bacteria living in soil and water, break-down NO_3, and use the energy released in the process, to support themselves. The process is called denitrification. Denitrifying bacteria are anaerobic because they grow in environments where there is little or no oxygen, such as sediments and deep layers in the soil near the water table, flooded soils, etc. The gaseous N_2 released as a byproducts then escapes and enters the atmosphere where from it came. Thus, the cycle is completed (Singh *et al.,* 2006).

3.3.1. Processes of Nitrogen Cycle

3.3.1.1. Nitrogen Fixation

The conversion of nitrogen from the atmosphere into a form readily available to plants and hence to animals and humans is an important step in the nitrogen cycle, that determines the supply of this essential nutrient. There are four ways to convert N_2 into more chemically reactive forms (Smile, 2000):

1. Biological fixation: some symbiotic bacteria (most often associated with leguminous plants) and some free-living bacteria are able to fix nitrogen and assimilate it as organic nitrogen. Examples of mutualistic nitrogen fixing bacteria are the Rhizobium bacteria, which live in legume root nodules. These species are diazotrophs. An example of the free-living bacteria is Azotobactor.
2. Industrial N-fixation: in the Haber-Bosch process, N_2 is converted together with hydrogen gas into ammonia which is used to make fertilizer and explosives.
3. Combustion of fossil fuels: automobile engines and thermal power plants, which release various nitrogen oxides (NOx).
4. Other processes: additionally, the formation of NO from N_2 and O_2 due to photons and especially lightning, are important for atmospheric chemistry, but not for terrestrial or aquatic nitrogen turnover.

3.3.1.2. Assimilation

Plants can absorb nitrate or ammonia ions from the soil via their root hairs. If nitrate is absorbed, it is first reduced to nitrite ions and then ammonium ions for incorporation into amino acids, nucleic acids, and chlorophyll (Smile, 2000). In plants which have a mutualistic relationship with rhizobia, some nitrogen is assimilated in the form of ammonium ions directly from nodules. Animals, fungi, and other heterotrophic organisms absorb nitrogen as amino acids, nucleotides and other small organic molecules.

3.3.1.3. Ammonification

When a plant or animal dies, or an animal excretes, the initial form of nitrogen is organic. Bacteria, or in some cases, fungi, converts the organic nitrogen within the remains back into ammonia, a process called ammonification or mineralization.

3.3.1.4. Nitrification

The conversion of ammonia to nitrates is performed primarily by soil-living bacteria and other nitrifying bacteria. The primary stage of nitrification, the oxidation of ammonia is performed by bacteria such as the Nitrosomonas species, which converts ammonia to nitritres. Other bacterial species, such as the Nitrobacter, are responsible for the oxidation of the nitrites into nitrates (Smile, 2000).

3.3.1.5. Denitrification

Denitrification is the reduction of nitrites back into the largely inert nitrogen gas, completing the nitrogen cycle. This process is performed by bacterial species such as Pseudomonas and Clostridium in anaerobic conditions (Smile, 2000). They use the nitrate as an electron acceptor in the place of oxygen during respiration. These faculatively anaerobic bacteria can also live in aerobic conditions.

3.3.1.6. Anaerobic Ammonium Oxidation

In this biological process, nitrite and ammonium are converted directly into dinitrogen gas. This process makes up a major proportion of dinitrogen conversion in the oceans.

3.3.2. Global Nitrogen Reservoir and Fluxes

The biological N_2 fixation on global basis is about 44-200 Tg (1 Tg = 1×10^{12} g) on land and 1-130 Tg in ocean, atmospheric fixation 0.5-30 Tg and industrial fixation 60 Tg yr^{-1}. The global nitrogen reservoirs and fluxes are given in Table 3.2.

3.3.2.1. Human Influences on the Nitrogen Cycle

During the past century, human activities clearly have accelerated the rate of nitrogen fixation on land, effectively doubling the annual transfer of nitrogen from the vast but unavailable atmospheric pool to the biologically available forms. The major sources of this enhanced supply include extensive cultivation of legumes, growing use of the chemical fertilizers, and pollution emitted by vehicles and industrial plants. Furthermore, human activity is also speeding up the release of nitrogen from long-term storage in soils and organic matter including 40% of the nitrous oxide, 80% or more of nitric oxide, and 70% of ammonia releases (Vitousek *et al.,* 1997) (Figure 3.3). N_2O has risen in the atmosphere as a result of agricultural fertilization, biomass burning, cattle and feedlots, and other industrial sources (Chapin *et al.,* 2002). N_2O has deleterious effects in the stratosphere, where it breaks down and acts as a catalyst in the destruction of atmospheric ozone. Ammonia in the atmosphere has tripled as the result of human activities. It is a reactant in the atmosphere, where it acts as an aerosol, decreasing air quality and clinging on to water droplets, eventually resulting in acid rain. Fossil fuel combustion has contributed to a 6 or 7 fold increase in NO_x actively alters atmospheric chemistry, and is a precursor of troposphere ozone production, which contributes to smog, acid rain and increases nitrogen inputs to ecosystems (Smile, 2000).

Table 3.2. Global nitrogen reservoirs and annual fluxes (based on Rosswall, 1983)

Reservoir	Atmosphere		Ocean		Land	
	N_2	39×10^5	Plants	0.30	Plants	11-14
	N_2O	1.4	Animals	0.17	Animals	0.2
	NH_3+ NH_4	174×10^{-5}	Microbes	0.02	Microbes	0.5
	$NO +$ NOx	6×10^{-4}	Dissolved detritus	530.0	Detritus	1.9-3.3
	NO_3	1×10^{-4}			Soil organic matter	300
	Org N	1×10^{-3}	Particulate detritus	3-240	Soil inorganic	160
			N_2 (dissolved)	22000		
			N_2O	0.2		
			NO_3	570		
			Nitrite	0.5		
			NH_4	7		

Table 3.2. Continued

Fluxes	
Biological N fixation	44-200 (land), 1-130 (ocean)
Atmospheric fixation (lighting)	0.5-30
Industrial fixation	60
Industrial combustion and fuel burning (NOx, N_2O	10-20
Fire	10-200
Biogenic NOx production	0-90
Denitrification	43-390 (land), 0-330 (ocean)
Denitrification (N_2O)	16-69 (land), 5-80 (ocean)
Nitrification N_2O production	4-10
Ammonia volatilization	36-250
Dry and wet NH_3/NH_4 deposition	110-240
Dry and wet NOx deposition	40-116
Dry and wet deposition of organic N	10-100
River runoff	13-40

Source: Singh *et al*., 2006

Values for reservoir are in petagrams (1 Pg= 1×10^{15}g) and those of fluxes are in teragrams nitrogen per year (1 Tg= 1×10^{12}).

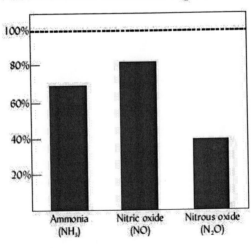

Figure 3.3. Human activities are responsible for a large proportion of the global emissions of nitrogen-containing trace gases (Vitousek *et al.*, 1997).

The global use of N fertilizers is expected to reach 250Tg N yr^{-1}. This may result in ammonia volatilization of 25 Tg yr^{-1}. If cattle population continues to grow at 2% yr^{-1}, about 40 Tg ammonia will be released per year into the environment through their waste, which enters the soil system as well as finds it way to hydrological system through leaching, groundwater flow and runoff. Increased coal consumption will amount to 25 Tg N yr^{-1}. Thus, ammonia flux from land to atmosphere in the coming decades may be as high as 100 Tg yr^{-1}. The rate of oxidized N deposition in terrestrial ecosystems of Asia from anthropogenic sources has changed from 1.1Tg N yr^{-1} in 1961 to 7.1 Tg N yr^{-1} in 2000 and is further expected to change to 15.5 Tg N yr^{-1} in the year 2030, while the rate of reduced N deposition has changed from 3.8 Tg in 1961 to 11.3 Tg N yr^{-1} in 2000; and is expected to change to 15.5 Tg N yr^{-1} in the year 2030 (Zheng et al., 2002).

3.4. PHOSPHORUS CYCLE

Phosphorus (P) is present in a small quantities in Earth's crust (0.09 wt %) and plays an essential role in biological systems. The cycling of phosphorus, a biocritical element in short supply in nature is an important earth system process. Variations in the phosphorus cycle have occurred in the past. For example, the rapid uplift of the Himalayan Tibet Plateau increased chemical weathering, which led to enhanced input of phosphorus to the oceans. This drove the late Miocene "biogenic bloom". Additionally, phosphorus is redistributed on glacial timescales, resulting from the loss of the substantial continental margin sink for reactive P during glacial sea-level low stands. The modern terrestrial phosphorus cycle is dominated by agriculture and human activity. The natural riverine load of phosphorus has doubled due to increased use of fertilizers, deforestation and soil loss, and sewage sources (Filippelli, 2008).

Fertilizer-based food production and the application of P, along with N, K, and other micronutrients, in commercially available fertilizers boomed after World War II. The use of fertilizer contributes substantially to the dissolved P cycle. Deforestation, typically by burning after selective tree harvesting, converts the standing stock of P in plant matter to ash. This P is rapidly dissolved, leached from the ash, and transported in rivers over timescales of a year or two (Schlesinger, 1997). The lack of roots destabilizes the landscape, resulting in loss of the O and A soil horizons, relatively rich in organic P, from many of these areas.

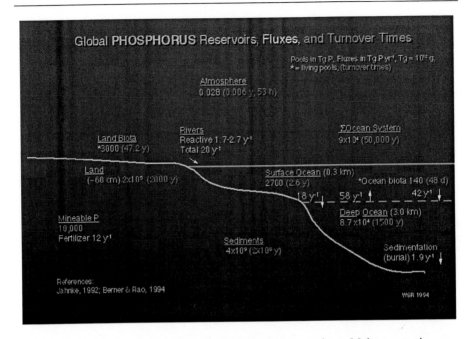

Figure 3.4. Global phosphorus reservoirs, fluxes and turnover times. Major reservoirs are nderlined; pool sizes and fluxes are given in Tg (10^{12} g) P and Tg P yr-1. Turnover times (reservoir divided by largest flux to or from reservoir) are in Parentheses. To convert Tg P to moles P, multiply by 3.2×10^{10}.

Fluxes of phosphorus involve plant-available forms such as dissolved phosphate ion (PO_4), dissolved organic P, and particulate inorganic P. In fact, most of the P in the biosphere exists as phosphate. In terrestrial ecosystems, biota (plants and animals), available P of soil, and humus and microbial P of soil, and in oceans dissolved P, biota, and detritus form tightly coupled compartments for P cycling (Figure 3.4).

The rocks and their deposits formed in the geological past constitute the major reservoir of phosphorus. The land biota contain only a small amount of P (about 2600 Tg P) compared to soil (more than 96000 Tg) or mineable rocks (19000 Tg) (Table 3.3). Similarly, although total dissolved P in the oceans is 80,000 Tg, marine biota contains only 50-120 Tg P. Further, although land plants have 20-50 times as much P as marine biomass, they cycle only 200 Tg P yr^{-1} compared to 600-1000 Tg yr^{-1} cycled by marine biota. As a result of erosion of soils and rocks, or leaching of dissolved forms, phosphates are released into rivers and lakes etc. Some of them are precipitated in lake sediments while the majority escapes into the sea, where part of it is deposited in the shallow sediments and part of it is lost to the deep sediments. As organisms die and sink to lower levels, surface waters become depleted of phosphate supplies, but upwelling of deep waters returns some phosphorus to the surface. Through the uplifting of sediments and the harvest of fish from sea the phosphorus returns from sea to land. According to one estimate, through the fish we eat, 60,000 tons of elementary phosphorus returns annually. Sea birds by depositing their faecal material on land also contribute significantly to the return of phosphorus to the cycle. But this is insufficient to compensate for the loss from the land to the sea (Singh *et al.*, 2006).

Table 3.3. Global phosphorus reservoirs (Tg P) and fluxes (Tg P yr^{-1}) (based on Richey, 1983)

Atmosphere		Land		Ocean	
Reservoirs					
Particulates		Biota	2.6×10^3	Biota	50-120
Over land	0.025	Soil	96×10^3-	Dissolved (inorganic)	80×10^3
			160×10^3		
Particulate					
Over Oceans	0.003	Mineable rock	19×10^3	Detritus (particulate)	650
		Fresh Water	90	Sediment	840×10^6
Fluxes					
Atmosphere (land) Over land	Atmosphere (ocean) Over Ocean	1.0	Marine (dissolved)	Biota	600-1000
Atmosphere Over Ocean	Atmosphere Over Land	0.3	Marine detritus	Sediment	2-13
Atmosphere	Land	3.2	Terrestrial Biota	Soil	200
Atmosphere	Ocean	1.4	Mineable rock	Soil	14
Land	Atmosphere	4.3	Soil	Freshwater	4-7
Ocean	Atmosphere	0.3	Freshwater (dissolved)	Ocean	1.5-4
			Freshwater (particulate)	Ocean	17

Source: Singh *et al.*, 2006.

In brief, the phosphorus cycle can be summarized as follows:

i.   Major reservoir of P is found in the lithosphere.
ii. From rock, phosphorus is released by the process of weathering.
iii. From weathering, the released phosphorus is transported to soil by wind or water as inorganic phosphate.
iv. Inorganic phosphorus is absorbed and assimilated by plants. In most soils, the amount of available phosphorus is about 0.01% of the total phosphorus in soil.
v.   From plant, phosphorus moves through the food chain in organic form.
vi. The debris of plant, animals and microorganisms return organic phosphates.
vii. The dead organic matter is acted upon by the phosphatising bacteria to release inorganic phosphorus from bound organic form.
viii.Loss of phosphorus occurs in runoff water in deep-ocean sediments.
ix. Phosphorus is returned from shallow marine deposits in fish harvest and guano deposits of fish-eating birds and geological uplift.
x.   Turnover of organic phosphorus occurs due to phosphatese activity associated with root activity and microbial populations.
xi. The precipitation of phosphorus in marine habitats limits primary productivity.
xii. Phosphorus limited lakes are oligotrophic.

3.4.1. Human Impact on Phosphorus Cycle

The human-era terrestrial P cycle is dominated by agriculture and human activity. Human have markedly increased the rate of phosphate mobilization by extracting phosphate fertilizers from mineable rocks. However, a major part of phosphorus added as fertilizer is immobilized in the soil because of abundant supplies of calcium, aluminium and iron which bind with it in the presence of oxygen. A portion of P is harvested in crops and is later released to water bodies through sewage and waste (Figure 3.5). In due course of time, P moves to sediment. Without massive use of phosphate fertilizers in the modern agriculture, yields can hardly be increased or sustained. So the use of fertilizers has continued to increase. The known phosphate reserves are expected to last for only 60-160 years at projected levels of use in future (Singh *et al.*, 2006).

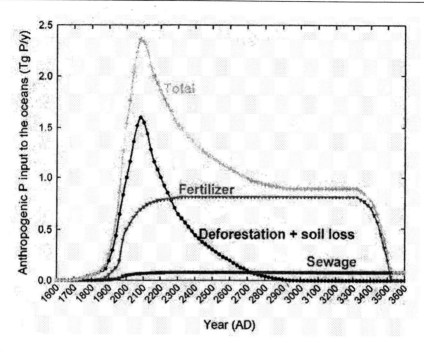

Figure 3.5. Reconstruction and projections of anthropogenic phosphorous delivered to the oceans as a result of several processes, including fertilization, deforestation and soil loss, and sewage. The projection is based on reports of population and arable land published by the World Health Organization and the Intergovernmental Panel on Climate Change. The fertilizer drop off coincides with the depletion of known P reserves. The total integrated anthropogenic input of P (1600-3600AD) is 1860 Tg (Filippelli, 2008).

3.5. THE SULPHUR CYCLE

Sulphur is one of the components that make up proteins and vitamins. Sulphur is important for the functioning of proteins and enzymes in plants, animals that depend upon plants for sulphur. Sulphur has a unique role in plant and animal metabolism because it is essential for the synthesis of amino acids such as cystein, methionine, cystein and co-enzymes-thiamine and biotin and certain vitamins. Sulphur group of amino acids determine the structure of proteins. The roots of green plants absorb sulphur from the soil mainly in the sulphate form, like nitrate and phosphate.

Most of the earth's sulphur is tied up in rocks and salts or buried deep in the ocean in oceanic sediments. The sulphur cycle has large reservoir in soil and sediments, and the smaller one in the atmosphere, and it does not cycle smoothly.

Reservoirs of sulphur are fossil fuels and elemental sulphur deposits in rocks and minerals. It is among the ten most abundant elements in the earth crust. The concentration of sulphur is low in the atmosphere, where it exists in gaseous form as H_2S and SO_2 and in particulate form as sulphates and sulphides. The combustion of fossil fuels releases SO_2 and other sulphur gases and solids into the atmosphere. Lithosphere contains about 24×10^9 Tg S, ocean water 1.3×10^9 Tg, soil (including land plants) 2.7×10^5 Tg and atmosphere only 4.8 Tg. The global reservoir and fluxes of sulphur are shown in Table 3.4 and the global sulphur cycle in Figure 3.6. It is clear that the main source of mobile sulphur is the lithosphere and its ultimate sink is the ocean, with limited recycling within the continental atmosphere-soil system. The sulphur flux from continental atmosphere to the oceanic atmosphere is 81 Tg S yr^{-1} against the opposite flux of 20 Tg S. This imbalance is on account of the anthropogenic sulphur emission to the continental atmosphere (Singh *et al.*, 2006).

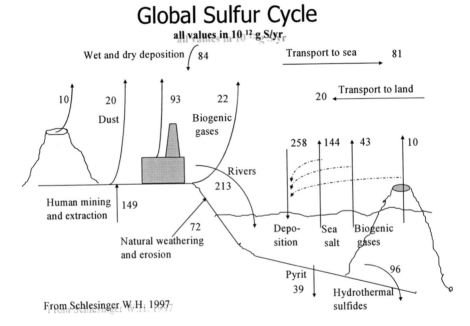

Figure 3.6. The global sulphur cycle.

Sulphur compounds play a major role for our environment and the climate system. On the one hand they contribute to acid rain. But they are also important for the formation of clouds. Finally, a lot of sulphur is brought into the air by volcanic eruptions. If it was a strong eruption, the emitted particles can go up to the stratosphere and cool down half our planet by 1-2°C.

Table 3.4. Global sulphur reservoirs (Tg S) and fluxes (Tg S yr^{-1})
(based on Frency *et al.*, 1983)

Atmosphere		Land		Ocean
Reservoirs				
Sulphate in aerosols	0.7	Plants	760	Oceanic water 1.3×10^9
SO$_2$	0.5	Soil	2.6×10^5	
Carbonyl sulphide	2.3	Soil Organic Matter	1.1×10^4	
Other reduced sulphur gases	0.8	Total	2.7×10^5	
Troposphere total	4.3	Lithosphere	24.3×10^9	
Stratosphere	0.5			
Total	4.8			
Fluxes				
Combustion, smelting etc.	113	Uptake of SO$_2$ by land surface and vegetation	17	
Volcanic gases	28	Wash out over land	51	
Aeolian emission	20	Dry deposition of SO$_2$ over land	16	
Biogenic gases (land)	16	Uptake of SO$_2$ by ocean	11	
Biogenic gases (coastal and sea)	20	Wash out over ocean	230	
Sea spray	140	Dry deposition of SO$_2$ over ocean	17	
Emission of long-lived reduced S compounds	5			

Source: Singh *et al.*, 2006.

Chapter 4

HYDROLOGICAL ANALYSIS

4.1. INTRODUCTION

Human disturbance of the water cycle is a global phenomenon affecting both river discharge and the transport and processing of sediments, carbon, and nutrients in aquatic ecosystems. This has led to significant changes in the transport of constituents from terrestrial biomes to the oceans with poorly known impacts on global biogeochemistry. Despite its enormous importance to human society, an arguably neglected element of the global change question surrounds the linkages between the continental land mass, river systems, and the coastal/near shore environment (IGBP Report No. 39).

The high mountains that are tectonically active yield most sediment, and about 70% of the global flux is produced in southern Asia and large Pacific Islands, an area dominated by mountainous source areas, rapidly increasing population and land-cover change (Milliman & Meade, 1983). The specific manner in which humans have modified these linkages requires further study. To address the full dimension of anthropogenic change as it relates to water, the need is to consider greenhouse-induced climate change, expanding management and conversion of land-use/cover, and an ubiquitous alteration of the water cycle for activities such as irrigation, flood control, hydro-electric production, industrial withdrawal, and waste processing.

The issue of riverine transports is therefore of direct relevance to the broader question of contemporary global change, with obvious policy implications. Anthropogenic effects such as impending climate change, stream regulation, eutrophication and increased erosion will collectively influence the long-term functionality of inland and coastal ecosystems to supply drinking water, support fisheries, and regulate floods and transport and process wastes. The reality that approximately 60% of the world's population resides in the coastal zone (IGBP-LOICZ, 1995) makes the issue of land-river-coastal zone linkages timely and relevant to global habitability.

Human activities can cause rapid modification of the characteristics of the landscape and its hydrological function. Individual and societal endeavours, such as agriculture, urbanization, deforestation, and drainage of wetlands, and their associated hydrological features continuously change the land surface. These changes not only have socio-economic and ecological consequences at different time scales; they may also have an impact on the climate and on the hydrological cycle, at least at the regional scale.

4.2. THE HYDROLOGIC CYCLE

Although the existence of the hydrologic cycle was recognised as early as third century B.C. by Theophrastus, the notion was accepted only in the seventeenth century after Pierre Perrault compared the rainfall and discharge of Paris Basin, and after Marriott confirmed that rainfall is the source of water flowing through streams and springs (Kotwichi, 1991).

The hydrologic cycle is the orderly sequence of movement of water from sea to land and back to sea (Figure 4.1). It is a recurring process and involves incessant movement of water in different states from oceans to the atmosphere (vapour), then to land (liquid or solid), and finally back to sea via streams and rivers (Kale & Gupta, 2001).

Water constitutes 70% weight of organisms; it is a medium of biological activity; it is a geological agent eroding soil from one place and depositing it in another; it carries nutrients in dissolved state and acts as an agent of their distribution; it is an agent of energy transfer and use; it is an important factor in controlling variations in temperature because it has the highest specific heat (1 cal/g/°C) and latent heat (80 cal/g heat is required to convert 1 g of water to ice, and 536 cal/g is required for converting 1 g of water at 100 °C to water vapour with no change in temperature) (Singh et al., 2006).

The water cycles rapidly but a majority of it (about 98.70%) is locked up in the lithosphere. The 0.13 geogram of water (1 geogram= 1×10^{20} g) which is the source of all precipitation, constitutes an infinitesimal proportion of the global water (266,069.88 $\times 10^{20}$). The atmospheric vapours are equivalent to 2.5 cm of rain, i.e., world's 9 days' supply and thus turn over every 9 days. The water in the rivers turns over every 12 days. It is estimated that by the year 2100 humans will have used or polluted all fresh water reserves, and by the year 2230 human may have to depend only on precipitation (National Commission on Agriculture, 1976).

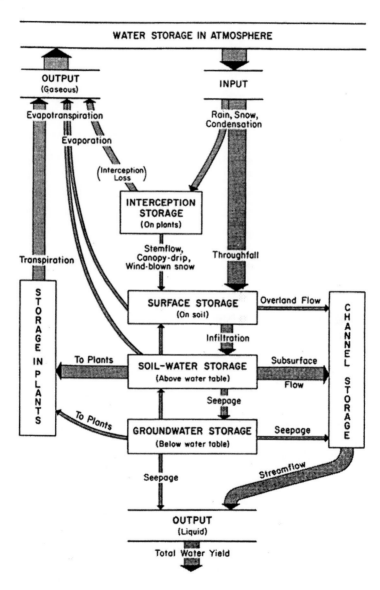

Figure 4.1. The hydrological cycle consists of a system of water storage compartments and the solid, liquid, or gaseous flows of water within and between the storage points (from Anderson *et al.,* 1976).

It is of interest to know how energy drives the hydrological cycle. For the uphill movement of the water via evaporation enormous amounts of solar energy are needed. To evaporate 1 g of water 0.536 kcal energy is required, total annual evaporation is 440×10^{18} g water, so 2.35×10^{20} kcal (i.e., $0.536 \times 440 \times 10^{18}$) of energy is spent annually for evaporation. This is one-fifth of the total energy income at the surface of the earth.

4.3. HYDROLOGICAL CHARACTERISTICS

India is endowed with a network of 2.89 million km^2 of river basin systems, with an average annual discharge of 1900 billion m^3 of water (Table 4.1). There are 14 major rivers, covering 2.35 million km^2 of basin area with an annual discharge of 1406 billion m^3 of water. Of this huge quantity of water, only 1.5 million km^2 of cultivable area can be irrigated due to the lack of technical and financial resources (Bobba *et al.*, 1997).

Table 4.1. Principal rivers of the Himalayan region- basic statistics

River Basin	Area (km²)	Mean discharge (m³/s)	Glacial melt in river flow (%)	Population ×1000	Population density	Water availability (m³/person/year)
Indus	1081718	5533	44.8	178483	165	978
Ganges	1016124	18691	9.1	407466	401	1447
Brahmputra	651335	19824	12.3	118543	182	5274
Irrawaddy	413710	13565	small	32683	79	13089
Salween	271914	1494	8.8	5982	22	7876
Mekong	805604	11048	6.6	57198	71	6091
Yangtze	1722193	34000	18.5	368549	214	2909
Yellow	944970	1365	1.3	147415	156	292
Tarim	1152448	146	40.2	8067	7	571
Total				1324386		

Source: IUCN/IWMI, Ramsar Convention and WRI, 2003; Mi & Xie, 2002; Chalise & Khanal, 2001; Merz, 2004.

The total annual volume of water discharged by all the river systems in India is 1645 billion m^3. Of this, the major river systems contribute 85%, while the medium and minor, including desert, rivers contribute 7% and 8%, respectively. The total flow in all the rivers of the world is estimated as 27,137 billion m^3, of which two thirds enters the sea and the rest enters lakes and swamps (Bobba *et al.*, 1997). The total flow of all the rivers in India is 6% of the flow of all rivers in the world (Rao, 1979).

Like other mountain areas, the Himalayan mountains are sometimes called the Roof of the World, is demonstrating a number of noticeable impacts related to global climate change, the most widely reported being rapid reduction in many glaciers which has implications for water resources. The region plays an important role in global atmospheric circulation, biological and cultural diversity, water resources, and the hydrological cycle, apart from the beauty of its landscape and provision of other ecosystem amenities (Bandyopadhyay & Gyawali, 1994). The region is the source of the nine largest rivers in Asia, the basins of which are home to over 1.3 billion people (Table 4.2). Environmental change in the HKH region affects much of inland China, Central and South Asia, and the mainland of Southeast Asia. The Himalayan region is the most critical region in the world in which melting glaciers will have a negative effect on water supplies in the next few decades (Barnett *et al.*, 2005).

Table 4.2. Hydrological characteristics of major river basins in India

River	Discharge (m³/s)	Runoff (m³/yr)	Drainage area (km²)	Mean basin elevation (m)	Mean river length (km)
Brahmputra	16,186	510,450	690,000	5000	2900
Ganges	14,556	459,040	970,000	3900	6315
Indus	6589	207,800	1,165,000	2500	2900
Godavari	3,330	105,000	290,000	400	1500
Krishna	2,146	67,675	260,000	420	1300
Mahanadi	2,113	66,640	132,090	500	885
Narmada	1,291	40,705	90,000	760	1300
Cauvery	664	20,950	72,000	630	765
Tapti	570	17,982	62,000	740	700
Minor rivers	3,000	104,736	240,00		

Source: Bobba *et al.*, 1997.

Large quantities of sediments leave the Himalaya through its rivers. The Himalayan Rivers rank amongst the top rivers in terms of suspended sediment load (Meybeck & Ragu, 1995) (Figure 4.2). In terms of suspended sediment delivery, the rivers originating from the Central Himalaya such as the Karnali,

Sethi Nadi, Tamur, Sun Khosi, Arun and Marsyangdi shows the highest figures, with values of more than 65 t ha^{-1} yr^{-1} (Lauterburg, 1993). The western Himalayan Rivers such as the Jhelum, Chenab and Indus have low sediment delivery rates of below 15 t ha^{-1} yr^{-1}.

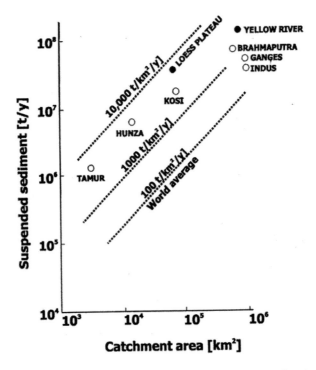

Figure 4.2. The sediment load of selected South Asian rivers compared to the global average (Ferguson 1984 in Alford 1992).

4.3.1. Himalayan Watershed Case Study

Himalaya has been divided into a large number of watersheds. Watershed is regarded as a functional unit in upland for natural resource management and development. River geochemistry varies considerably with the type of lithology, topography, land-use/cover change. Understanding this variability in relation to changes in land-use and hydrology will be critical in the prediction of nutrient budgets for the functioning of watersheds. Therefore, a representative watershed has been selected in the Sikkim Himalaya for analysis.

4.3.1.1. Stream Discharge, Sediment Concentration, Soil and Nutrient Loss

The land-use change and hydrology of Mamlay watershed have been studied to determine the effect of land-use on sediment and nutrient flux from the system and its role in climate change. The selected watershed is part of the catchment of the Rangit River, the second largest river of the Sikkim Himalaya. The watershed has a total area of 3014 ha and the total stream length is 82.6 km. All the streams attain significant sizes during the rainy season. Seasonal streams dry up by January-May in the watershed. The highest discharge of 4143 l s^{-1} was recorded in the rainy season in 1999 followed by 4137 l s^{-1} in 2000 and the lowest of 850 l s^{-1} and 840 l s^{-1} in summer season, in the respective years, in the Rinjikhola, the outlet of the watershed. For the different streams the discharge was in order Pockcheykhola > Sombareykhola > Chemcheykhola > Tirikhola > Rangrangkhola and the significant variation was observed only in rainy season. The discharge in various streams showed high seasonality and direct relationship with precipitation. Most of the precipitation was received in the monsoon and consequently discharge was highest in this season. Analysis of variance showed that streams, season and stream×season varied significantly ($P<0.0001$) (Sharma & Rai, 2004).

The sediment concentration varied distinctly with seasons in different streams and the outlet of the watershed. The sediment concentration during 1999 and 2000 ranged from 9-61 mg l^{-1} in winter, 8-59 mg l^{-1} in summer, and 14-399 mg l^{-1} in the rainy season. Analysis of variance showed significant variation between streams and seasons and its interaction was also significant ($P<0.0001$). The highest sediment concentration in the rainy season was mainly because of high precipitation and extensive agricultural practices followed in this season. Seasonal and yearly soil loss value was recorded in stream waters of the micro-watersheds and total watershed for the two year period 1999-2000 are presented in Table 4.3. The soil loss from different micro-watersheds ranged from 0.001-7.48 t ha^{-1} in 1999 and 0.001-6.62 t ha^{-1} in 2000. The soil loss rate from the total watershed ranged between 6 to 7 t ha^{-1} yr^{-1} during the two years of the study. The total soil loss from the watershed with an area of 30.14 km^2 is significant, ranging from 18295 t yr^{-1} in 1999 to 21953 t yr^{-1} in 2000 (Sharma & Rai, 2004).

Nutrient loss from soil sediment in the stream water of the micro-watersheds and total watershed is given in Table 4.4. The total nitrogen loss ranged between 0.76 and 18.79 kg ha^{-1} yr^{-1} for the micro-watersheds, and was about 33 kg ha^{-1} yr^{-1} as an average over two years at the watershed outlet. Organic carbon loss ranged from 9 to 162 kg ha^{-1} yr^{-1} in the micro-watersheds, while the average annual loss from the outlet of the watershed was 267 kg ha^{-1}. Similarly, the total phosphorus loss ranged from 0.21 to 2.86 kg ha^{-1} yr^{-1} in the micro-watersheds, but the average annual loss from the outlet of the watershed was 5.39 kg ha^{-1} yr^{-1}. Total nitrogen loss was 100 t yr^{-1}, organic carbon was 804 t yr^{-1} and total phosphorus 16 t yr^{-1} from the entire watershed area of 30.14 km^2 (Rai & Sharma, 1998b).

Table 4.3. Seasonal and yearly soil loss (tons) estimated using discharge and sediment concentration of micro and total watershed

Parameter		Micro-watersheds					Total Water-shed
		Pokchey khola	Chem-chey khola	Tiri khola	Sombaray khola	Rang-rang khola	Rinji khola[*]
Area (ha)		788	717	509	635	365	3014
1999	Winter	48	47	284	10	0.03	744
	Summer	13	30	33	0.01	0.00	87
	Rainy	1235	854	3491	977	0.44	17464
	Total	1296	931	3808	987	0.47	18295
	Soil loss (t ha^{-1})	1.64	1.30	7.48	1.55	0.001	6.07
2000	Winter	33	43	98	7	0.01	500
	Summer	10	19	28	0.01	0.00	83
	Rainy	1118	793	3242	948	0.4	21370
	Total	1161	855	3368	955	0.41	21953
	Soil loss (t ha^{-1})	1.47	1.19	6.62	1.50	0.001	7.28

[*]Watershed outlet.
Source: Sharma & Rai, 2004.

Across the micro-watersheds and total watershed, the carbon loss through runoff and sediment was analyzed. Organic carbon loss through sediments ranged from 0.014 to 136 t yr^{-1} in micro-watershed, while the annual loss from the outlet of the watershed was 833 t yr^{-1} (Table 4.5). The loss of soluble carbon through runoff water ranged between 0.96 to 814 t yr^{-1} for the micro-watersheds and was about 2025 t yr^{-1} at the watershed outlet. Streamflow concentrations of soluble carbon showed the most distinct seasonal trend. On seasonal basis highest loss was recorded during rainy season (Table 4.6) and it varied significantly

Table 4.4. Nutrient loss (in kg) estimated from soil sediment in stream waters of micro-watersheds and total watershed

Micro-watershed	Season	1994-1995			1995-1996		
		Total nitrogen	Organic carbon	Total phosphorus	Total nitrogen	Organic carbon	Total phosphorus
Pokche-ykhola	Winter	236.04	1345.40	32.90	124.60	1319.50	23.28
	Summer	0.17	0.96	0.02	149.52	1583.40	27.93
	Rainy	1315.08	7495.80	183.30	4069.08	43091.10	760.09
	Total	1551.29	8842.16	216.22	4343.20	45994.00	811.96
	Nutrient loss (kg ha-1)	1.97	11.22	0.27	5.51	58.36	1.03
Chemche-ykhola	Winter	160.17	912.95	22.33	238.52	2525.90	44.56
	Summer	0.42	2.40	0.06	188.68	1998.10	35.25
	Rainy	918.87	5237.45	128.08	3453.20	36569.00	645.05
	Total	1079.46	6152.80	150.47	3880.40	41093.00	724.86
	Nutrient loss (kg ha-1)	1.51	8.58	0.21	5.41	57.31	1.01
Tirikhola	Winter	177.03	1009.05	24.68	338.20	3581.50	63.18
	Summer	3.71	21.14	0.52	170.88	1809.60	31.92
	Rainy	9382.59	53479.65	1307.78	7290.88	77209.60	13611.92
	Total	9563.33	54509.84	1332.98	7799.96	82600.70	1457.02
	Nutrient loss (kg ha-1)	18.79	107.09	2.62	15.32	162.28	2.86

Table 4.4. Continued

Micro-watershed	Season	1994-1995			1995-1996		
		Total nitrogen	Organic carbon	Total phosphorus	Total nitrogen	Organic carbon	Total phosphorus
Sombare-ykhola	Winter	261.33	1489.55	36.43	49.84	527.80	9.31
	Summer	0.84	4.80	0.12	42.72	452.40	7.98
	Rainy	6255.06	35653.10	871.85	388.04	4109.30	72.49
	Total	6517.23	37147.45	908.40	480.60	5089.50	89.78
	Nutrient loss (kg ha-1)	10.26	58.50	1.43	0.76	8.01	0.14
Rinji-khola*	Winter	3287.70	18739.50	458.25	2232.12	23637.90	416.96
	Summer	118.02	67.27	16.45	1733.72	18359.90	232.86
	Rainy	102854.43	586258.05	14336.18	90673.20	960219.00	16937.55
	Total	106260.15	605064.82	14810.88	94639.04	1002216.80	17678.37
	Nutrient loss (kg ha-1)	35.26	200.75	4.91	31.40	332.52	5.87

* Watershed outlet; Source: Rai & Sharma, 1998b.

(*P*<0.0001). All the streams showed highest concentration in the rainy season. On micro-watershed wise, mean yearly stream water soluble carbon concentration for the two years study period was recorded highest at Pockcheykhola (814 t yr^{-1}) compared to mean concentration values with other micro-watersheds (Sharma & Rai, 2004).

Table 4.5. Carbon loss (tons) through sediment in different streams water of Mamlay watershed

Season	Micro-watersheds					Total watershed
	Pockhey khola	Chemchey khola	Tiri khola	Sombaray khola	Rangrang khola	Rinji khola[*]
Winter	1.2	1.6	3.9	0.2	0.003	17.8
Summer	0.4	0.7	1.1	0.0003	0.00	3.0
Rainy	45	28	131	34	0.014	812
Total (t yr^{-1})	46.6	30.3	136	34.2	0.0143	833

[*] Watershed outlet.
ANOVA: Streams $F_{4, 30}$ = 427.65, *P*<0.0001; Season $F_{2, 30}$ = 1769.40, *P*<0.0001. Streams x Season $F_{8, 30}$ = 382.901, *P*<0.0001.
Source: Sharma & Rai, 2004.

The discharge in various streams showed high seasonality and direct relationship with precipitation. About 90% of annual precipitation was received in the monsoon and the discharge was highest in this period. Many streams dried completely during the summer season mainly in the mid hills because of deforestation and extensive human activities. This belt is located between two major thrusts of the watershed where water percolates from upper thrust and appears in the lower thrust through sub-surface flow. Sediment concentration also showed seasonality similar to discharge. The sediment concentration in different seasons at all streams showed direct relationship with precipitation. The highest sediment concentration in rainy season was attributed to (i) high rainfall during this period, (ii) steep slopes and (iii) extensive cultivation of the soil practices in this season. The soil loss rate from the total watershed ranged from 6 to 7 t ha^{-1} yr^{-1} during the two years of study. Rawat & Rawat (1994) reported about 2 t ha^{-1} yr^{-1} soil loss in a normal rainfall year from a watershed in the Central Himalaya where the rainfall is comparatively low. The two year average of the annual sediment flux from the watershed was 667 t km^{-2} yr^{-1}. This is within the range of 500-1000 t km^{-2} yr^{-1} reported for the Himalayan region by Milliman & Meade (1983). Soil loss as high as 3005 t km^{-2} yr^{-1}, was recorded in an agro-ecosystem less than 5 years of shifting cultivation (Toky & Ramakrishnan, 1981).

Table 4.6. Soluble carbon (tons) loss through different streams water of Mamlay watershed

Season	Micro-watersheds					Total watershed
	Pokchey khola	Chemchey khola	Tiri khola	Sombaray khola	Rangrang khola	Rinji khola[*]
Winter	74	77	74	14	0.06	51
Summer	7	14	16	0.2	0.00	148
Rainy	733	559	388	401	0.9	1826
Total (t yr^{-1})	814	650	478	415	0.96	2025

[*] Watershed outlet.
ANOVA: Streams $F_{4, 30}$ = 5650.89, $P<0.0001$; Season $F_{2, 30}$ = 46064.92, $P<0.0001$. Streams x Season $F_{8, 30}$ = 3980.46, $P<0.0001$.
Source: Sharma & Rai, 2004.

Mean annual estimates of organic carbon export via stream flow differ somewhat with micro-watersheds. Carbon in soluble form was lost more through runoff than sediment movement. The higher concentration values at stream water are related to the mean annual discharge.

4.3.1.2. Precipitation, Overland Flow and Soil Loss

Precipitation was recorded at two locations representing different slope and aspects in the watershed covering sub-tropical and temperate belts for the period of two years from 1999-2000. The average annual precipitation for the two years period was 2992 mm in temperate belt and 1295 mm in sub-tropical belt of the watershed. Overland flow (percentage of rainfall during rainy season) was recorded to be highest in open cropped area sub-tropical (10.86%) and lowest in cardamom based agroforestry (2.80%) (Table 4.7) (Sharma & Rai, 2004). Usually the non-forested sites had a greater overland flow of water compared with adjacent forested and agroforestry sites. Overland flow was a function of the size of the rain-shower. Nevertheless, the magnitude of overland flow was too small to play a significant role in the wider context of flooding. The overland flows involve subsurface systems and that most of the water is transmitted to streams by lateral down slope flow within the soil.

Rainfall was distributed seasonally and more than 90% was received during May to October. The overland flow in the open cropped area was highest because of intensive cultivation and steep aspect of the land. It takes between 200 and 1000 years to form 2.5 cm of top soil under cropland conditions, and even longer under forest conditions (Pimental et al., 1995). About 80% of the World's agricultural land suffers moderate to severe erosion, and 10% suffers slight to moderate erosion (Speth, 1994). Croplands are most susceptible to erosion because their soil is repeatedly tilled and left without a protective cover of vegetation. A survey of agricultural fields on unterraced slopes showed more than 60% pebbles/stones. Participatory inventory with farmers also revealed high soil erosion problem and the indicators as observed by farmers were exposure of red soil and stones of deeper soil profile. The overland flow decreased in mandarin orange based agroforestry as a result of protection by trees. In the sub-tropical forest relatively high amount of overland flow was recorded because of high biotic pressure. Prior to the year of experimentation this forest was totally in degraded condition and devoid of ground vegetation and understory species. This has contributed to greater overland flow and soil loss. Overland flow and soil loss from the wasteland was low compared to open cropped area as it was not disturbed and was covered by ground vegetation. Similar observations on fallow lands were also made in shifting agriculture system in North-Eastern India by Toky & Ramakrishnan (1981). According to an estimate made by Shah (1982) nearly 85% of all agricultural land already suffers from severe erosion problems. The overland flow and the soil loss in large cardamom based agroforestry system were lower because of good tree canopy and under-story thick large cardamom bush coverage. Temperate natural forest dense showed relatively lower overland flow and soil loss. In the Central Himalaya, comparatively less overland flow was recorded from the temperate forest but the soil loss was more than the temperate forest of the present study (Pandey et al., 1983; Singh et al., 1983; Negi et al., 1998). The high overland flow in the present study located in the eastern Himalaya was the consequence of higher rain intensity and more annual precipitation. In spite of more overland flow in the temperate natural forest in the present study, soil loss was less than the Central Himalaya because of complete ground vegetation, thicker forest floor litter and more stratification of the forest. Large cardamom based agroforestry is a traditional practice of the region and is regarded to be profitable and sustainable farming system. The less overland flow values in temperate natural forest and large cardamom based agroforestry indicate that the catchment areas under these land-uses encourage high infiltration and subsurface flow. Bren & Turner (1979) and Bren (1980) studied the surface runoff on steep forested infiltrating slopes in Australia and reported that overland flow was very low (0.005% of the rainfall). The hydrological response of a forested hill slope to rain is often dominated by the lateral down slope movement of water within the soil system. Overland flow may be a rare occurrence on such forested watershed.

Table 4.7. Overland flow (% of rainfall) and soil loss (kg ha^{-1}) in different land-use/cover of Mamlay watershed

Land-use	Overland flow	Soil loss
Temperate natural forest dense	2.57	16
Temperate natural forest open	3.92	22
Subtropical natural forest open	4.56	27
Cardamom based agroforestry system	2.80	18
Mandarin based agroforestry system	4.77	31
Open cropped area temperate	10.25	480
Open cropped area subtropical	10.86	525
Wasteland area temperate	3.78	24
Wasteland area subtropical	3.90	25

ANOVA: Overland flow – Land-use $F_{8, 18}$ = 104.14, $P<0.0001$; Soil loss – Land-use $F_{8, 18}$ = 1131.52, $P<0.0001$.
Source: Sharma & Rai, 2004.

In most areas, raindrop splash and sheet erosion are the dominant forms of erosion. Erosion is intensified on sloping land, where more than half of the soil contained in the splashes is carried downhill. The sediment movement from the temperate natural forest dense was 16 kg ha^{-1} during the monsoon period and this is 17 times lower than the values recorded from open cropped area of the sub-tropical belt (Table 4.5). There is a dramatic rise in sediment output from the landslide and newly constructed road sites consequent to the formation of channels. Erosion increased dramatically on steep cropland. Soil loss was recorded highest in open cropped area sub-tropical (525 kg ha^{-1}) when compared to forests and agroforestry systems.

Total organic carbon concentration in parent soil and eroded soil was estimated during the rainy season in different land-use/covers. Concentration of total organic carbon content was higher in eroded soil than the parent soil. Total organic carbon content in the parent soil upto 30 cm depth ranged from 10 to 26 mg g^{-1}, highest being recorded in temperate natural forest dense and very little variation was observed in other land-use/covers. But the highest organic carbon concentration in eroded soil was recorded in wasteland area temperate (40.8 mg g^{-1}) and lowest from temperate natural forest dense (32.2 mg g^{-1}). An ANOVA test on organic carbon concentration between eroded and parent soils and between land-uses shared a statistically significant variation ($P< 0.0001$) (Sharma & Rai, 2004).

Soil and organic carbon losses from open cropped area was more than 90% of total watershed indicating that agriculture practice without agroforestry in such unterraced sloping land and in high rainfall areas are highly vulnerable. Therefore, strong agroforestry based agriculture such as mandarin, cardamom and horti-agri-silvi system is recommended in the watershed for conservation of soil, water and nutrients in such a fragile upland farming system. Reliable and proven soil conservation technologies include ridge planting, no-till cultivation, crop rotations, strip cropping, grass-strips, mulches, living mulches, terracing, contour planting, cover crops and windbreaks (Pimental et al., 1995). Although the specific processes vary, all conservation methods reduce erosion rates by maintaining a protective vegetative cover over the soil, which is often accompanied by a reduction in the frequency of ploughing. Ridge planting, for example, reduces the need for frequent tillage and also leaves vegetative cover on the soil surface year round, and crop rotations ensure that some part of the land is continually covered with vegetation. Each conservation method may be used separately or in combination to control soil erosion. To determine the most advantageous combination of appropriate conservation technologies, the soil type, specific crop and climate (rainfall, temperature and wind intensity), as well as the socioeconomic conditions of the people living in a particular site must be considered.

4.4. BIOGEOCHEMISTRY OF HIMALAYAN RIVERS

Local hydrology of every river in the world is likely to be affected by land-use and climate change. In Himalaya, major river basins may have already been affected by climate change. Climate change affects different aspects of local hydrology of river such as timing of water availability and quantity, as well as its quality. Changes in river hydrology will induce risks to water resources facilities that includes flooding, landslides, and sedimentation from more intense precipitation events (particularly during the monsoon) and greater unreliability of dry season flows that possesses potentially serious risks to water and energy supplies in the lean season.

Eighty per cent of the sediments delivered to the world's oceans each year come from Asian Rivers and amongst these the Himalayan Rivers are the major contributors (Stoddart, 1969). The Himalaya contributes 500-1000 t Km^{-2} yr^{-1} of sediment (Milliman & Meade, 1983) and Sikkim in the north-eastern region as also shows similar value of 794 t km^{-2} yr^{-1} (Sharma, *et al.*, 2001). It is estimated that the river Brahmputra alone carries a suspended load of 800 million tons, the average sediment-yield from its catchments being of the order of 26000 ha meters (Raina *et al.*, 1980). River systems are also the major means of transport of dissolved materials, including inorganic nutrients and contaminants, which depend on riverine geochemical and hydrological processes. It is estimated that current rates of erosion are five times as great as the rates prevailing in the geological past (Singh *et al.*, 1983). The problem of erosion is still more acute in the high-sesmicity areas, which are geo-dynamically sensitive thrust areas.

Soil erosion is a major environmental and agricultural problem worldwide. Although erosion has occurred throughout the history of agriculture, it has intensified in recent years. The loss of soil degrades arable land and eventually renders it unproductive. Worldwide, about 12×10^6 ha of arable land are destroyed and abandoned annually because of unsustainable farming practices and only about 1.5×10^9 ha of land are cultivated (Pimental *et al.*, 1995).

Hydrologically linked ecosystems interact through the flow of nutrients through water. Nutrients discharged from uplands pass through low lands on their way to the sea. Understanding the dynamics of such nutrient flows requires knowledge of the effect of land-use on nutrient discharge and of the effects of uphill ecosystems on downhill ecosystems. The relationship between land-use/cover change and soil erosion and hydro-ecological process is a key element in understanding the little known local, regional, and global biospheric disruptions.

The long-term variations in the earth's atmospheric CO_2 levels are driven by a variety of geologic processes. CO_2 is added to the atmospheric reservoir by global volcanic degassing, the main source term; and the two major processes representing as sink: the uptake of CO_2 during rock weathering and that during photosynthesis and transformation to organic matter. The CO_2 removal via organic matter formation and rock weathering (mainly carbonate and silicate weathering) is eventually converted into dissolved organic carbon (DOC); particulate organic carbon (POC) and dissolved inorganic carbon (DIC as HCO_3^- ion) pools of the rivers; which are subsequently exported to the ocean. It is well accepted that changing the atmospheric CO_2 consumption by continental weathering can have an influence on earth's climate.

4.4.1. River Transport Data Analysis

One way to study the rate of rock weathering is to evaluate the chemical composition of river water. For Ca-Mg silicates, the riverine flux of dissolved HCO_3^- is a direct measure of the uptake of atmospheric CO_2 owing to weathering. Such an approach is important to estimate the present-day inorganic and organic denudations of the continents and their associated CO_2 consumption. The mountainous regions are dominated by rapid mechanical erosion, in turn, increases the surface area of fresh minerals available for chemical attack. The chemical breakdown of detrital material in mountainous regions is enhanced by the abundance of easily weathered sedimentary silicates and orographic rainfall on mountain slopes which flush away mechanical and chemical erosion products, thereby constantly exposing fresh minerals. These effects are most marked in the Himalaya, under the influence of greater rainfall associated with the Asian monsoon. Data from the eight largest rivers draining the Himalayan-Tibetan region (Ganga, Brahmputra, Chang Jiang, Xijiang, Irrawady, Indus, Mekong, and Huanghe) show that almost 25% of the total dissolved load reaching the ocean today comes a watershed area that represents nearly 5% of the earth's land surface. Thus, a disproportionate fraction of the earth's chemical weathering occurs in this small region in Asia. It is likely that the late Cenozoic uplift of the Tibentan plateau would have resulted in regionally and hence globally, higher chemical erosion rates, causing a drawdown of atmospheric CO_2 and global cooling (Sarin, 2001).

The contribution from rock weathering (silicates, carbonates and evaporate rock), rain and atmospheric inputs dominates the dissolved load of rivers. On a global scale, relative proportions of solutes derived from carbonate weathering account for almost 70% of the dissolved load. Based on the recent compilation of the data from 60 largest rivers, Gaillardet *et al.*, (1999) have summarized that 300 $\times 10^6$, 640×10^6 and 140×10^6 t yr^{-1} of dissolved salts are derived from weathering of silicates, carbonates, and evaporates, respectively; and that the total flux of dissolved material from continental chemical weathering is 1080×10^6 t yr^{-1}. Assuming that nearly 51% of the surface runoff is represented by these 60 largest rivers, Gaillardet *et al.*, (1999) have further estimated that about 2130 $\times 10^6$ t yr^{-1} of dissolved material is derived from chemical weathering, of which about 26% (550×10^6 t yr^{-1}) is contributed by silicate weathering.

The present day average flux of atmospheric/soil CO_2 consumed by global chemical erosion and exported to the oceans as bicarbonate ion is estimated to be 0.28-0.30 Gt C yr^{-1} (Sarin, 1999). The amount of CO_2 consumed by silicate weathering for the river drainage basins lying in the Himalayan-Tibetan region, ranges from (54 to 832) $\times$ 10^9 mol yr^{-1}; with enhanced consumption rates for Irrawady (832 $\times$ 10^9 mol yr^{-1}) and that for Ganga (471 $\times$ 10^9 mol yr^{-1}). On a global scale, the CO_2 consumption for silicate rock weathering is 8.7 $\times$ 10^{12} mol yr^{-1} (0.148 Gt C yr^{-1}) for carbonate weathering (Gaillarder *et al.*, 1999). In addition, the recent estimates show that young volcanic rocks consume 3 $\times$ 10^{12} mol C yr^{-1}. Thus, the present day CO_2 flux consumed by chemical weathering of silicate rocks is 11.7 $\times$ 10^{12} mol yr^{-1} (0.140 Gt C yr^{-1}) of which nearly one-sixth is consumed by weathering processes in the Himalaya (Sarin, 2001).

The specific CO_2 consumption rates for the Himalayan-Tibetan region range from (146 to 2250) $\times$ 10^3 mol km^{-2} yr^{-1}; with highest consumption rate for the Irrawady and lowest for the Indus. Overall, the specific CO_2 consumption rates associated with the weathering of both silicate and carbonate rocks in the Himalayan-Tibetan rivers (700 $\times$ 10^3 mol km^{-2} yr^{-1}) are nearly three times higher than the world average river (250 $\times$ 10^3 mol km^{-2} yr^{-1}) and four to five times higher than those in the Amazon basin. Such higher CO_2 consumption rates have a potential to affect the earth's biogeochemical cycle of C in the present day scenario of growing anthropogenic CO_2 emission (Sarin, 2001).

Chapter 5

THE GLOBAL CLIMATE CHANGE

5.1. INTRODUCTION

The earth is the only planet in our solar system that supports life. The complex process of evolution occurred on Earth only because of some unique environmental conditions that were present: water, an oxygen rich atmosphere, and a suitable surface temperature. Only the earth has an atmosphere of the proper depth and chemical composition. About 30% of incoming energy from the sun is reflected back to space while the rest reaches the earth, warming the air, oceans, and land, and maintaining an average surface temperature of about 15˚C.

Global climate change occurs naturally and periodically and is often attributed to continental drift, variations in the earth's axis and orbit, variations in solar energy output and the frequency of volcanic activity. The average surface temperature of the earth has been increasing since the end of the Little Ice Age (15th-18th centuries) (Bajracharya *et al.*, 2007). Over the past few decades, the global climate has been undergoing significant changes due to increasing human population and its activities, especially during the last century. Some of the important changes that are occurring in the global climate relate to increased resource consumption; depletion of fossil fuel reserves that were generated over several hundred million of years in the geological past; and the large scale changes in land-use and land-cover. Excessive use of terrestrial productivity and accessible fresh water, to meet the demand of food, fodder and fibre of the ever increasing human population is having large impact on the natural systems. So much so that another Geological Era, called the Anthropocene Era (Box 5.1) (Singh *et al.*, 2006; IGBP, 2001).

The average surface temperature of the earth has increased between 0.3°C and 0.6°C over the past hundred years and the increase in global temperature is predicted to continue rising during the 21st century. On the Indian sub-continent, temperatures are predicted to increase between 3.5 and 5.5°C by 2100 (IPCC, 2001a) and an even greater increase is predicted for the Tibetan Plateau (Lal, 2002). It is estimated that a 1°C rise in temperature will cause alpine glaciers worldwide to shrink as much as 40% in area and more than 50% in volume as compared to 1850 (IPCC, 2001b; CSE, 2002).

Box 5.1. Anthropocene Era

- Concentrations of climatically important gases have substantially increased.
- Coastal and marine habitats dramatically altered; 50% mangroves removed and wetland reduced by one-half.
- 22% recognised marine fisheries depleted and 44% at their limit of exploitation.
- Extinction rates increasing sharply.
- 40% of known oil reserves exhausted by humans in last 150 years that took hundreds of millions years to generate.
- 50% land surface transformed affecting biodiversity, soil biology and climate.
- More nitrogen is fixed synthetically for fertilizers than is fixed naturally.
- >50% of all accessible fresh water is appropriated for human use and groundwater resources are being rapidly depleted.

The most significant changes brought by human activities are the increase in concentration of carbon dioxide and other greenhouse gases in the troposphere, and depletion of stratospheric ozone. The naturally occurring greenhouse gases in the lower atmosphere regulate the global temperature, whereas ozone in the stratosphere filters out the harmful ultraviolet radiations reaching the earth's surface. Thus, any change of ozone in the stratosphere or gain in greenhouse gases in the lower atmosphere may lead to global climate change (Box 5.2).

This chapter discuss about the sources and the increasing concentration of greenhouse gases and the effects of climate change on terrestrial ecosystems.

Box 5.2. What is Climate Change

Climate change refers to a statistically significant variation in either the mean state of the climate or in its variability, persisting for an extended period (typically decades or longer). Climate change may be due to natural internal processes or external forcing, or to persistent anthropogenic changes in the composition of the atmosphere or in land use.

5.2. GREENHOUSE EFFECTS

Short wave radiation from the sun (0.2 to 0.4µm) reaches the earth's surface unhindered but the outgoing long wave radiation radiated from the earth (4.0 to 100 µm) is partially retained by the atmosphere. The energy that is radiated from the earth passes through the atmosphere to outer space through the region of atmospheric window (approximately 10% of the energy). The atmosphere absorbs a substantial portion of the long wave radiation emitted by the earth (approximately 90% of energy radiated from earth), and radiates energy back to it. This downward flux of the long wave radiation, called greenhouse flux, keeps the earth warm. Thus, the phenomenon of trapping and reradiating of heat by greenhouse gases in the atmosphere is referred to as greenhouse effect (Box 5.3) (Singh *et al.*, 2006). The gases that most actively absorb the radiant heat energy are known as greenhouse gases or radiatively active gases. Nitrogen and oxygen are most abundant gases in the atmosphere accounting for about 78% and 20.9% of the total gaseous volume. The remaining 1% is composed of argon, water vapour, carbon dioxide, ozone, and other gases. Water vapour, CO_2, N_2O, CH_4, CFCs and ozone occur in minute quantities in the atmosphere but play a critical role in maintaining even temperatures on the earth; these are the radiatively active greenhouse gases. The excessive increase in concentrations of greenhouse gases in the atmosphere will retain more infrared radiation, resulting in enhanced greenhouse effect. The consequent increase in the global mean temperature is referred to as *Global Warming*. Recently, the occurrence of extreme weather or climatic events has been used to indicate overall climate change. The average temperature of the earth's surface did not vary much between 1940 and 1970 AD, but a continuous rise in temperature has been recorded since 1970. Over the past few decades, human activity has significantly altered the atmospheric composition, leading to climate change of an unprecedented character.

Box 5.3. Greenhouse Effect

Greenhouse gases act to warm the atmosphere and enhance the greenhouse effect. CO_2 is the most important greenhouse gas, but methane (CH_4), chlorofluorocarbons (CFCs), tropospheric ozone, and nitrous oxide (N_2O), when taken together, are as important as CO_2.

5.3. EVIDENCE FOR GLOBAL WARMING

Evidence of the warming is present in many different forms. The most obvious evidence is in the temperature records. Analysis of temperature data indicates rapidly rising temperatures over the past 140 years. There are many temperature-dependent phenomena that also indicate the earth is warming.

- Earth's mountain glaciers are melting
- Antarctica's ice sheets are breaking up
- Sea level is rising
- The temperature of the global ocean is rising
- Northern hemisphere permafrost is melting
- Arctic pack ice is thinning and retreating
- The tree line in mountain ranges is moving upward
- Many tropical diseases are spreading toward the poles and to higher elevations in the tropics.

5.4. GREENHOUSE GASES

Land-use/cover change directly affects the exchange of greenhouse gases between terrestrial ecosystems and the atmosphere. Carbon dioxide (CO_2), CH_4, CFC-11, CFC-12 and N_2O are the most important greenhouse gases. Among these, CH_4 is the only gas which directly affects tropospheric chemistry and controls numerous chemical processes and constituents in the troposphere and stratosphere (Crutzen, 1994). The greenhouse gases are increasing due to industrial development, increase in energy production and use, land-use/ cover changes and the intensive agricultural practices. The change in land-use is a complex phenomenon having its regional impact on climate and weather conditions. For example, the conversion of a forest to grazing land or a cropland is a local activity, causing loss of carbon to the atmosphere. However, the loss of soil organic matter has global effects on the carbon cycle and the greenhouse gases. Similarly, biomass burning associated with agricultural practices in Southeast Asia is an important regional activity, but has global environmental impacts due to the production of carbon monoxide, which is linked to increasing methane concentration in the troposphere.

Greenhouse gases reduce the emission of thermal radiation to space, thereby affecting the energy budget everywhere around the globe, i.e., warming the surface. The sources and trends of increase in the concentration of greenhouse gases in the atmosphere are summarized in Table 5.1. The global atmospheric concentration of CO_2 has increased from a pre-industrial value of about 280 ppm to 379 ppm in 2005. The atmospheric concentration of CO_2 in 2005 exceeded by far the natural range (180 to 300 ppm) over the last 650,000 years as determined from ice cores. The annual CO_2 concentration rate was greater during the last 10 years (1995-2005 average: 1.9 ppm per year) than it has been since the beginning of continuous atmospheric measurements (1960-2005 average 1.4 ppm per year) although growth rates vary from year to year (IPCC, 2007).

Table 5.1. Increase in the concentrations of greenhouse gases in the atmosphere as affected by human activities

Greenhouse Gases	Pre-industrial 1750 AD Concentration	Concentration in 2005 AD	Increase since 1750AD	Atmospheric Life Time (years)
Carbon dioxide (CO_2)	280 ppm	379 ppm	35%	5-200
Methane (CH_4)	700 ppb	1774 ppb	151%	12
Nitrous oxide (N_2O)	270 ppb	319 ppb	17%	114
Chlorofluorocarbons (CFC-11 +	0	282 ppt		45 (CFC-11) 260 (HFC-
Hydrofluorocarbon (HFC-23)				23)

ppm-parts per million, ppb-parts per billion, ppt-parts per trillion.

5.4.1. Carbon Dioxide

The increase in CO_2 concentration in the atmosphere has been largely the result of anthropogenic activities i.e., fossil fuel burning, deforestation and changes in land-use/cover. The main cause of increase in atmospheric carbon dioxide in recent years has been combustion of fossil fuels (coal, oil etc.) which release about 21 billion tons of CO_2 in the atmosphere annually. Currently, the industrial or developed nations account for about 60% of these emissions, however, contributions from developing nations are also growing and are likely to increase more than that of developing countries in future. Country-wise CO_2 emissions are given in Table (5.2). Presently, per capita per year emission is about 20 t in USA, 16 t Canada, 10.4 t Russia, 10 t in Europe, 9.2 t Japan, 7.7 t South Africa, 2.4 t in China, 1.8 t Brazil and 0.9 t in India. Thus, US having 5% of global population accounts for release of 22% CO_2 in the atmosphere. Of the total sources of CO_2, biotic portion accounts for about 20-30% whereas land-use/cover related to soil-borne sources contribute 10-15%. The other known major source of carbon dioxide is deforestation which is estimated to account for 15-30% of annual carbon dioxide emission. Every year about 0.8% of the total global forest area is cleared for other use. Forest clearing releases 0.3-2.6 Gt ($1Gt= 10^9$ t) of carbon annually as CO_2 (Singh *et al.*, 2006).

There is a clear evidence of changes having occurred in the composition of greenhouse gases in the lower atmosphere during the last century (IPCC, 2001) and over the long time scales of glacial and interglacial periods (Petit *et al.*, 1999). The analysis of the air trapped in the ice showed that CO_2 concentration has varied between 180 to 280 ppm (280 CO_2 molecules per million molecules of air) in the past 160,000 to 420,000 years during the glacial and interglacial periods (Barnola *et al.*, 1987; Petit *et al.*, 1999). The CO_2 concentration in the atmosphere varied in parallel with temperatures; the concentration being higher when temperatures were high. The concentrations of CO_2 were about 25% lower during the glacial periods than during the preindustrial interglacial periods. Thus, the CO_2 levels in the air currently are higher than at any time in the past 420,000 years. The atmospheric concentration of carbon dioxide over the past 1000 years is shown in Figure 5.1.

In recent times, atmospheric concentration of carbon dioxide has increased from preindustrial levels of 280 ppm to about 379 ppm in 2005. This trend will continue, if humans continue to burn fossil fuels and clear forests at current rate, particularly in tropics.

5.4.1.1. Carbon Losses from Land-use/Cover Change

Substantial carbon is transferred from the terrestrial reservoir to the atmosphere as a result of land-use/cover change. Changes in land-use through anthropogenic and climatic forcing result in vegetation conversion, which is likely to influence the balance of storage and flux of carbon in terrestrial ecosystems (Global Change Report No. 4, 1988). Estimating shifts of carbon due to land-use/cover change is a key process in determining impacts of disturbance on C storage in ecosystems. Allocation includes the pool of C in biomass components, and the fluxes of C between atmosphere, plants and soil. Possible adverse consequences of land transformation have drawn attention to the inventory and dynamics of biospheric carbon (Chan, 1982). Land-use/cover change may contribute to the loss of soil carbon by changing the balance between biomass production and decomposition. Intensive cultivation can also decrease soil carbon, contributing to terrestrial net fluxes of carbon to the atmosphere and decreased net primary productivity (Burke *et al.*, 1989; Johnson, 1992).

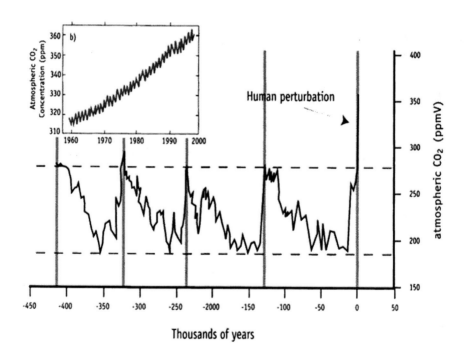

Figure 5.1. Atmospheric CO_2 concentration from Vostok ice core and Hawaii-no-analogue state, with the human perturbation superimposed (based on Keeling & Whorf, 2000).

Deforestation Occurred in Europe over many centuries as the population increased. In North America a similar deforestation occurred following colonization by Europeans (McNeill, 2000). Deforestation of the humid tropics is a relatively recent phenomenon, and is of great concern because its scale is likely to influence the global climate system, as well as threaten the rich diversity of species and the lives of indigenous people. In many tropical countries, development plans required substantial deforestation (Laurance *et al.*, 2001a&b)

and it is likely that tropical deforestation will continue into the foreseeable future, ceasing only when people realize the heritage value of the forest, as has happened in Europe.

Agriculture Organization (FAO), giving deforestation rate of 6.4 million hectares per year, equivalent to 0.55% of the total tropical forest per year. However, the often-quoted data from Dixon et al., (1994) suggests 15.4 million hectares per year. Assuming the average carbon content of rain forests to be 152 tC ha^{-1} the FAO figure implies a transfer of carbon to the atmosphere of only 0.97 Gt C yr^{-1} whilst Dixon's figure suggests 2.34 Gt C yr^{-1}. For much of the period in the 1980s and 1990s, Brazil's deforestation rate was 20,000 km^2 yr^{-1}, starting from a forest with an initial area of about 4 million km^2. Assuming a carbon density of 152 tC ha^{-1} this implies a flux of 0.3 Gt C yr^{-1}.

Table 5.2. Country wise growth rates of CO_2 emissions, 1990-2004

Country	CO_2 emissions (million tons)	Growth Rate (1990-2004)	CO_2 emissions per capita (tons)
USA	6046	25	20.6
China	5007	109	3.8
Russia	1524	-23	10.6
India	1342	97	1.2
Japan	1257	17	9.9
Germany	808	-18	9.8
Canada	639	54	20.0
UK	587	1	9.8
Korea	465	93	9.7
Italy	450	15	7.8
World	28983	28	4.5

Source: The Times of India, November, 2007.

5.4.1.2. Himalayan Watershed Carbon Loss

Area-weighted standing crop values for vegetation, litter, humus and soil are calculated on each land-use/cover type in the entire watershed and presented in Table 5.3. Total vegetation C in forested land-use ranged between 30.6 and 84.6×10^3 tC. In the agroforestry this value ranged from 0.01 to 4.91×10^3 tC. The overall range of soil C was 45.6 to 215×10^3 tC in forested land area, between 2.61 to 29×10^3 tC in agroforestry systems, between 8.5 to 40×10^3 tC in wastelands and between 15 to 24×10^3 tC in open cropped areas of subtropical and temperate belts. Total stand carbon in the studied watershed area (3014 ha) was 624×10^3 tC, total

C stored in the soil to a depth of 1 m was 456×10^3 t. Total vegetation C was 161×10^3 t, litter C 5.33×10^3 t and humus C 1.44×10^3 t in the whole watershed (Rai & Sharma, 2004).

Table 5.3. Area-weighted total stand carbon in the Mamlay watershed. Values are in (x 10^3 t C) (from Rai & Sharma, 2004)

Land-use/cover	Vegetation	Litter	Humus	Soil	Total Stand
Temperate natural forest dense	30.6	0.73	0.22	75.52	107.07
Temperate natural forest open	84.6	2.89	0.86	215.11	303.46
Subtropical natural forest open	32.7	1.1	0.23	45.6	79.63
Cardamom based agroforestry system	4.91	0.60	0.13	29.3	34.94
Mandarin based agroforestry system	0.01	0.007	-	2.61	2.63
Open cropped area temperate	3.84	-	-	15.30	19.14
Open cropped area subtropical	4.15	-	-	24.30	28.45
Wasteland area temperate	-	-	-	40.00	40.00
Wasteland area subtropical	-	-	-	8.52	8.52
Total watershed	160.81	5.33	1.44	456.26	623.84

The release of carbon or its accumulation depends on the standing stock of carbon in vegetation and soils and on the rates of deforestation. Land-use change detection study involving a total area of 3014 ha indicated changes in 1046 ha. This involved changes in vegetation stock (as dense forest converted into open forest to open cropped area to wastelands) and consequently in the standing crop of carbon. The total release of carbon to the atmosphere from the watershed was 305×10^3 t over a period of 13 years (1988-2001). Reductions in the biomass of the forest as a result of conversion were responsible for a net loss of 119×10^3 t vegetation C and 183×10^3 t soil C (Table 5.4). This translates into release of 7.78 tC ha^{-1} yr^{-1} from the entire watershed due to land-cover change (Rai & Sharma, 2004). Based on the results obtained for Mamlay watershed and assuming the same conditions, the total release of carbon from the entire Sikkim state and Indian Himalayan region can be assessed. Sikkim state occupies 284,779 ha forest land (about 40% of the total geographical area); land-use change (harvest and forest clearings) releases 22.16×10^5 t annually. If the same result is applied to the entire Indian Himalayan forests area (6.692 million ha), the total release of carbon would be 520×10^5 tC annually (Sharma & Rai, 2007). It is clear that because of overexploitation and continuous land conversion, the land-use/cover changes have become a net source of C to the atmosphere.

5.4.2. Methane

The global mean concentration of CH_4 in 2005 was 1774 ppb. Methane concentration has more than doubled since its concentration of 700 ppb during the preindustrial times (Table 5.1). Methane is very effective in causing warming as it absorbs infrared radiation of different wavelength than CO_2. On a molecule for molecule basis, methane is 21 times more powerful than CO_2 at trapping heat in the atmosphere. During the past 100 years, the concentration of CH_4 has increased 100% and over the geologic time scale its concentration has shown variations corresponding to that of atmospheric CO_2 (Singh et al., 2006). An estimated 15% of anticipated atmospheric warming is due to CH_4 (Houghton et al., 1992).

The total annual global emission of methane is estimated to be 420-620 Tg yr^{-1} (Khalil & Rasmussen, 1990), 70-80% of which is of biogenic origin (Bouwman, 1990). Emission sources of CH_4 are mostly biological and associated with human activities like landfills, livestock ruminants, and anaerobic respiration in soils of wetlands, and rice cultivation, and biomass burning. Rice paddies are considered to be among the most important sources of atmospheric CH_4, contributing from 60-100 Tg yr^{-1} (Watson et al., 1990). Methane emission from wetland rice agriculture can account for as much as 26% of the global anthropogenic methane budget (Neue & Roger, 1993). Fresh water wetlands produce CH_4 due to incomplete decomposition of organic matter in oxygen poor environments. Methane is also released due to anaerobic activity of methanogens in flooded rice fields, bogs, tundra, wetlands and lakes. Increase in methane (about 20%) is also related to production and use of fossil fuels and coal mining.

5.4.3. Chlorofluorocarbons

CFCs are considered to be most dangerous to planet's stability. The troposheric concentrations of chlorofluorocarbons have increased dramatically over the past 50 years (Walker et al., 2000). Automobile air conditioners use

CFC-12. This is among the most damaging of the CFCs. Automobiles are the largest single source of harmful CFCs. They make up about 27% of the CFCs released into the environment. The other major sources of CFCs are leaking air conditioners, refrigeration units and cleaning of electronic components, production of plastic foams and propellants in aerosol spray cans. The growth rate of CFCs in the atmosphere has been about 4% per year; having a long residence time of 22 to 102 years in atmosphere (Singh *et al.*, 2006). On a molecule for molecule basis, CFCs are 12000-15000 times more effective than CO_2 at trapping infrared radiation.

5.4.4. Nitrous Oxide (N2O)

Nitrous oxide is increasing in the atmosphere and causes as much as 5 to 6% of the anthropogenic greenhouse effect. The annual rate of increase of N_2O is 0.2 to 0.3% yr^{-1} and it has a long residence time of 150 yrs in the atmosphere. The main sources of N_2O are agricultural activity, biomass burning and industrial processes. For example, N_2O is produced during nylon production, burning of nitrogen rich fuels, denitrification of nitrogen rich fertilizers in soil and nitrate-contaminated groundwater. Deforestation and biomass burning are also significant sources of N_2O to the atmosphere. N_2O is 310 times more powerful than CO_2 at trapping heat in the atmosphere (Singh *et al.*, 2006).

An increase of greenhouse gases in the atmosphere leads to retaining of more infrared radiations and acts like a positive radiative forcing. CO_2 contributes about 60% of total warming, the contribution of CH_4, CFCs and N_2O being 20 and 14 and 6%, respectively. The relative contribution of different sources to greenhouse gases are as follows: burning of fossil fuel, 49% followed by industrial processes 24%, deforestation 14% and agricultural activities 13%, respectively.

5.5. POSSIBLE EFFECTS OF GLOBAL WARMING

It is true that the increase in concentration of greenhouse gases in the atmosphere would have profound impact on the climate of earth, including global warming, disturbed precipitation pattern and increased storm intensities (Box 5.4 & 5.5).

Box 5.4. Projected Impact of Climate Change

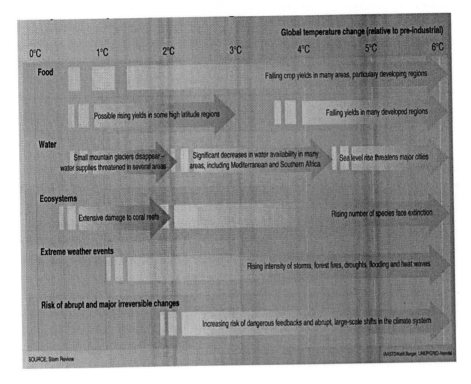

Box 5.5. Potential Effects of Global Warming

1. Elevated temperatures of the biosphere
   - melting of polar ice
   - increase in sea level
   - increase of methane from permafrost
2. Weather extremes
   - more rainfall during shorter periods
   - more evaporation and soil moisture deficiencies
3. Ecosystem disruption
   - stress and death of vegetation
   - shifting of species
4. Human Health
   - heat stress
   - migration of disease vectors

The predicted global warming would cause rise in sea level and shift in the range of distribution and phenology of organisms. Developed countries' share of world population 15%, but share of CO_2 emission is 50%. Temperature increase by 3-4°C would cause displacement of 330 million people due to floods, malaria infection for 220-400 million people due to floods, extinction of 20-30% of all the land species and between 2000 and 2004, about 262 million people affected by natural calamities. Of these 98% were in developing countries (The Times of India, November, 2007).

5.5.1. Weather and Climate

Since 1958 CO_2 is being measured at the Mauna Loa Observatory in Hawaii. It has the longest continuous record of the atmospheric CO_2 concentration and its seasonal and annual trends (Keeling & Whorf, 2005). There is a seasonal fluctuation in CO_2 of about 5 ppm during the year. The measurements have shown that compared to CO_2 concentration of 280 ppm in the early 19th century, the level has increased to 379 ppm in the year 2005. If the present rate continues, the concentration may be between 490 and 1260 ppm in the year 2100. The increase in concentration of carbon dioxide and other greenhouse gases causes the earth's average air temperature to increase. The year 1880 marks the beginning of the historical record for which there is enough data to provide credible information. During the last century, the global mean temperature has increased by approximately 0.6°C, reaching the highest level of 0.19°C during 1979-1998 (IPCC, 2001). It is expected that the average temperature of the earths near surface will increase by 1.4 to 5.8°C by the year 2100, assuming the level of greenhouse emission at 1990 level (IPCC, 2001). Most of the increase took place in the 1920s and 1990s. The 1990s were the warmest decade on record. For instance, 1998 was the warmest single year on record (Box 5.6).

This increase in the average temperature may not be uniform and vary over different places on earth. The IPCC report predicts that globally, average precipitation, particularly winter precipitation will increase in future in the northern hemisphere. Further, the frequency of extreme events (e.g., drought, floods, etc.) will substantially increase. This climate change may also be reflected in the glacial environment; some measurements indicate that Himalayan glaciers have been retreating at an increased rate since 1970 (Bajracharya et al., 2006). Rapid glacial melt can cause serious flood damage in heavily populated regions of Himalaya, whereas increase in floods from Asian monsoons has implications for the low lying areas.

Box 5.6.Global Temperature Data (from Oliver & Hidore, 2003)

- 1998 was the warmest year of record
- Seven of the 10 warmest years of record occurred since 1990
- The 1990s were the warmest decade on record
- 1980s were the second warmest decade on record
- The 10 warmest years of record occurred since 1983
- The mean temperature of earth increased about 0.4^0C in the 20th century
- The 20th century was the warmest century of the millennium

Westerlies have been disrupted already. This might explain why India's winter rains were poor this year; but may deliver a drenching. With 168 mm of rainfall, Delhi had its wettest May on record. In Uttar Pradesh, two storms killed 120 people. With seasonal rivers and sporadic rains, India's ecological miracle would become an ecological calamity.

5.5.1.1. Climate Change in the Himalayan Mountains

Climate change is a major concern in the Himalaya because of potential impacts on the economy, ecology and environment of the Himalaya and areas downstream. The Himalayan region, including the Tibetan Plateau, has shown consistent trends in overall warming during the past 100 years (Yao *et al.*, 2006). Various studies suggest that warming in the Himalaya has been much greater than the global average of 0.74°C over the last 100 years (IPCC, 2007; Du *et al.*, 2004). The increase in temperature over the Tibetan Plateau during the period from 1955-1996 is about 0.16°C per decade for the annual mean and 0.32°C per decade for the winter mean, and these exceed the increases in the northern hemisphere and for the same latitudinal zone in the same period. Furthermore, there is a tendency for the warming trend to increase with elevation on the Tibetan Plateau and in its surrounding areas. This suggests that the Tibetan Plateau is one of the most sensitive areas in terms of response to global climate change (Liu & Chen, 2000).

5.5.2. Melting of Polar Ice Caps and Glaciers

In high altitude areas of the mountain, an increased annual average temperature will cause increased thawing of permafrost and ice, including glaciers. It is estimated that the arctic ice sheet has shrunk by about 6% between

1978 to 1996. The arctic Greenland ice sheet (containing about 8% of world's ice) has thinned more than a meter per year on the average since 1993 along its southern and eastern regions. In Antarctica, there is massive ice cover (2.3 km thick), which represents 91% of earth's ice; the land-based ice sheets may not show much change.

Outside the poles, most ice melt has occurred in mountain and sub polar glaciers. Large scale ice melt would raise sea level and flood the coastal areas. As mountain glaciers recede, large regions that rely on the glacial runoff for water supply could experience severe water shortage. The change in glacier ice or snowmelt impacts water storage and the water yield to downstream areas. Sustained glacier retreat will cause two effects on river hydrology. First, large increases in river peak flows will increase the large-scale damage to downstream river valley schemes such as agriculture and water supply. Second, increasing threats arise from the formation and eventual outburst of high altitude glacial lakes. These climate changes will have a significant impact on the lives and property of downstream communities (Bjaracharya *et al.*, 2007).

Overall, the evidence supporting the phenomenon has been conclusive enough to make glacial melting and retreat an important indicator for climate change. The Himalayan glaciers have retreated by approximately a kilometre since the Little Ice Age (Mool *et al.*, 2001). Results show that recession rates have increased with rising temperatures. Evidence also shows that temperature changes are more pronounced at higher altitudes.

5.5.2.1. Retreat of Himalayan Glaciers

Himalayan glaciers cover about three million ha, or 17% of the global mountain area. They are the largest bodies of ice outside the polar caps. The total area of the Himalayan glaciers is 35110 km^2. The total ice reserve of these glaciers is 3735 km^3, which is equivalent to 3250 km^3 of fresh water. The Himalaya, the water tower of the world, is the source of nine great river systems of Asia: the Indus, Ganges, Brahmputra, Irrawaddy, Salween, Mekong, Yangtze, Yellow and Tarim, and are the water life line for 500 million inhabitants of the region, or about 10% of the total regional human population (IPCC, 2007).

Many studies have been carried out on the fluctuation of glaciers in the Indian Himalaya and significant changes have been recorded in the last three decades. The retreat of selected glaciers is summarised in Table 5.5; most of these glaciers have been retreating discontinuously since the post glacial period (Table 5.5). For example, the Siachen and Pindari Glaciers retreated at a rate of 31.5m and 23.5m per year, respectively (Vohra, 1981). The Gangotri glacier retreated by 15m per year from 1935 to 1976 and 23 m per year from 1985 to 2001 (Vohra, 1981;

Suresh C. Rai

Thakur *et al.,* 1991; Hasnain *et al.,* 2004). On average, the Gangotri Glaciers is retreating at a rate of 18m per year (Thakur *et al.,* 1991). Jeff Kargel of the USGS showed that the position of the Gangotri Glacier snout retreated about 2 km in the period from 1780 AD to 2001 (Figure 5.2) and is continuing to retreat. Shukla & Siddiqui (1999) monitored the Milam Glacier in the Kumaon Himalaya and estimated that the ice retreated at an average rate of 9.1m per year between 1901 and 1997. Dobhal *et al.,* (1999) monitored the shifting of the snout of the Dokriani Bamak Glacier in the Garhwal Himalaya and found that it had retreated 586m between 1962 and 1997. The average retreat was 16.5m per year. Matny (2000) found that the Dokriani Bamak Glacier retreated by 20m in 1998, compared to an average retreat of 16.5m over the previous thirty-five year. Table 5.6 shows the average retreat of other important glaciers in the Indian Himalaya. This indicates that the global warming has affected the snow-glacier melt and runoff patterns in the Himalaya. One of the best examples of glacier retreat is shown in Figure 5.2.

Table 5.4. Vegetation C, soil C, litter C, humus C, and total C injected into the atmosphere due to land-use/cover change during 1988-2001 (from Rai & Sharma, 2004)

From	To	Changed area (ha)	Release of vegetation $C (\times10^3$ t)	Release of litter $C (\times10^3$ t)	Release of humus $C (\times10^3$ t)	Release of soil $C (\times10^3$ t)	Total release $(\times10^3$ t)
Temperate natural forest dense	Temperate natural forest open	446.810	54.761	0.728	0.237	113.043	168.769
Temperate natural forest open	Open cropped area temperate	169.730	16.481	0.499	0.148	30.891	48.019
Temperate natural forest open	Wasteland area temperate	109.930	11.140	0.323	0.096	14.291	25.850
Subtropical natural forest open	Open cropped area subtropical	316.37	36.151	0.959	0.196	24.677	61.983
Subtropical natural forest open	Wasteland area subtropical	2.920	0.334	0.009	0.002	0.006	0.351
Total		1045.76	118.867	2.518	0.679	182.908	304.972

No changed in area observed in cardamom and mandarin based agroforestry systems.

Table 5.5. Retreat of some important glaciers in the Indian Himalaya

Glacier	Location	Period	Avg. Retreat rate (m/yr^{-1})	Reference
Siachen	Siachen		31.5	
Milam		1849-1957	12.5	Vohra (1981)
Pindari	Uttarakhand	1845-1966	23.5	
Gangotri		1935-1976	15	
Gangotri		1985-2001	23	Hasnain et al., (2004)
Bada Shigri	Himachal Pradesh	1890-1906	20	
Kolhani		1857-1909	15	Mayekwski & Jeschke (1979)
Kolhani	Jammu and Kashmir	1912-1961	16	
Machoi		1906-1957	8.1	Tiwari (1972) cited in WWF (2005)
Chota Shigri	Himachal Pradesh	1970-1989	7.5	Surendra et al., (1994)

Source: Bajracharya et al., (2007). Reprinted by permission.

Table 5.6. Average retreat rates of some major glaciers in the Indian Himalaya

Glacier Name	Retreat rate (m/yr^{-1})	References
Gangotri	18	Thakur et al., (1991)
Milam	9.1	Shukla & Siddiqui (1999)
Dokriani Bamak	16.7	Dobhal (1999)
	20 in 1998	Matny (2000)
Gara, Gor Garang, Shaune Garang, Nagpo Tokpo	4.2-6.8	Vohra (1981)
Bada Shigri, Chhota Shigri, Miyar, Hamtah, Nagpo Tokpo, Triloknath, Sonapani	6.8 for Chhota Shigri 29.8 for Bada Shigri	Srivastava (2003)
Janapa	20.5	
Jorya Garang	12.5	
Naradu Garang	16.2	Kulkarni (2004)
Bilare Bange	2.6	
Karu Garang	23.5	
Baspa Bamak	11.2	
Shaune Garang	40.5	Philip & Sah (2004)

Source: Bajracharya et al., (2007). Reprinted by permission.

Figure 5.2. Retreat of Gangotri glacier snout during the last 220 years from 1780 AD to 2001 (from Bajracharya *et al.,* 2007, reprinted by permission).

If present retreat trends continue, the total glacier area in Himalaya will likely shrink from the present 500,000 to 100,000 km^2 by the year 2035. Chinese experts predict that by 2050 the icy area on their side of the Himalaya will have shrunk by more than a quarter since 1950. Predictions for the Indian side are gloomier still. The East Rathang glacier in lofty Sikkim has shrunk by 2.5 km, or half its length, in decade. The Sikkim has 84 glaciers, some of which are melting rapidly. With glacial lake outburst floods likely to become frequent, 14 of the 266 glacial lakes in Sikkim are potentially dangerous. The shrinking of glaciers will restrict the water supply of Yangtze and Yellow river of China and Brahmputra, Indus and Ganges of India, affecting crops.

5.5.3. Changes in Plant and Animal Life

During the last three decades, there has been considerable scientific effort to determine plant responses to rising atmospheric CO_2. Under the elevated concentration of atmospheric CO_2, plants are able to increase their rate of photosynthesis. The CO_2 enrichment studies in greenhouses, growth chambers, and open-top chambers have suggested that growth of many plants could increase by about 30% on average, with a doubling of the atmospheric CO_2 concentration, in short-term, under favourable conditions of water, nutrients, light and temperature. Elevated CO_2 had a positive effect on ecosystem primary production as aboveground primary production increased by 19%, belowground primary production by 32% and net primary production by 12% (Nowak *et al.*, 2004). CO_2-enriched world of the future, the elevated CO_2 will be affecting the growth, reproductive effort, transpiration and stomatal conductance and chemical composition of wild and agricultural plants. India's agriculture will suffer more than any other country's. Assuming a global temperature increase of 4.4°C over cultivated areas by 2080, India's agricultural output is projected to fall by 30-40%.

There are also detectable changes in Earth's animal life. The population of Adelie penguins on the Antarctic dropped 40% in the last quarter of the 20th century. Off the coast of California, the sea surface has warmed as much as 1.5°C since 1951. This has led to an 80% decline of zooplankton, which is a basic component of the food chain (Oliver & Hidore, 2003). Coral reefs are one of the richest ecosystems on the planet, and they are dying at an unprecedented rate. Rates of natural migration and adaptation of species and plant communities appear to be much slower than the rate at which climate is changing. Thus, the population size and geographical range of many species may change as temperature and rainfall change.

Animal and plants have a specific range of distribution. With increasing global warming many species are expected to shift slowly from lower latitude to higher latitudes and from lower altitudes to higher altitudes. In the northern hemisphere, an increase of 500 m in the altitude means a decrease of about 3°C in mean temperature. So if temperature were to increase by 3°C, species distribution may shift by about 500 m up a mountain side. Similarly, the northern limits of species ranges are expected to move pole-ward by about 500 km. There are greater chances of extinction of local populations along range boundaries at higher latitudes or higher elevations. These changes in the distribution of species are likely to have marked effect on species diversity and ecosystem functioning of coral reefs, mountain ecosystems, coastal wetlands, tundra, boreal and temperate forests. The warming trends during the last 30 years or so have shown that the spring activities of organisms, like breeding and singing of birds, arrival of migrant birds, migration of butterflies and flowering in plants, have advanced a few days in terrestrial ecosystems especially in middle and higher latitudes (Singh et al., 2006). There are clear trends of northward and upward elevation shifts in range boundaries of birds, butterflies, and alpine herbs in response to climate change trends (Parmesan & Yohe, 2003).

5.5.4. Impact on Soil and Nutrient Cycling

The soil is the major pool of organic carbon. Soil carbon has global significance in carbon cycle, as it is a source of increase in CO_2 and CH_4 in atmosphere. Jenkinson et al., (1991) have analyzed the effect of global warming on the release of CO_2 from soils. They predicted that a global temperature rise of 0.3°C per decade will release an extra 61×10^{15} g CO_2 from the soil into the atmosphere by the year 2050. This will account for 19% of the total CO_2 release from fossil fuel combustion. Thus, a change in the soil organic carbon content will affect the CO_2 concentration in the atmosphere. The climate change scenario suggests changes in soils and vegetation in the boreal and tundra regions. A temperature rise will have positive effects on various soil processes like mineral weathering, decomposition of organic matter and soil biological processes. The rising atmospheric CO_2 could alter ecosystem carbon balance through positive and negative feedbacks to climate (Pendall et al., 2004).

5.5.5. Sea Level Change

Sea level is expected to rise as a result of global warming. From 1955 to 1995, the World ocean warmed by an average of 0.06°C from the surface down to about 3000 m. The Indian Ocean warmed 0.3°C down to a depth of 800m in a period of 20 years. Third IPCC report indicates a sea level rise of 1.5 ± 0.5 mm yr^{-1} for the 20th century. Several factors contribute to the sea level change that include the thermal expansion of ocean as it warms, the melting of glaciers, changes in the mass of Antarctica and Greenland ice sheets and changes in terrestrial water storage. Over the past century, melting of polar ice caps and increased mountain water flow has contributed on the average about 20% to the estimated 10-25 cm global sea level rise during the 20th century; rest being contributed by thermal expansion of oceans (Figure 5.3). It is estimated that sea level may rise by an additional 9 to 88 cm on the average by the year 2100, having large effects on coastal cities (IPCC, 2001). People living near coast would face danger from rising seas. The danger would be from occasional severe storms that would cause sudden flooding farther inland.

Rising sea levels will claim tropical islands of India i.e., Bay of Bengal have submerged two islands in the Sunderbans, where tigers roam through mangrove forests in the Ganges river delta, and a dozen more islands are under threat. A six-year study of the impact of future climate change on the world natural heritage site that India shares with Bangladesh came up with alarming results. Around seven million people are projected to be displaced due to submersion of parts of Mumbai and Chennai, if global temperatures were to rise by a mere 2°C. Many villages in Meghalaya will also submerge along neighbouring low-lying Bangladesh. Upto 18% of Bangladesh and many parts of coastal India will be submerged by 2050.

Global warming is bound to hit Indian economy and environment badly (Box 5.7). India could lose as much as 9% of its GDP, largely from events like submergence of low-lying coastal areas. The direct impact of this change in climate is seen to be far worse. About 7.1 million people in India would be affected by submergence of coastal lands if temperatures rise by 2°C. The economic loss, quite naturally would be high in urban zones such as Mumbai and Chennai. Mumbai alone could suffer loses up to $48 billion due to projected submergence.

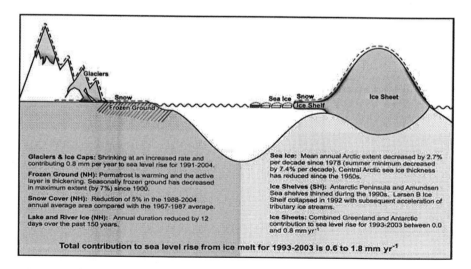

Figure 5.3. Total contributions to sea level rise from ice melt for 1993-2003 (from IPCC, 2007, reprinted with permission).

Box 5.7. The Cost for India

- At one-meter sea level rise would lead to submergence of 576,400 ha of land in India. This would displace approximately 7.1 million people.
- Economic impact of climate change on Mumbai could be US$48 billion (Rs 2,28,700 crore), while smaller cities like Balasore could lose US$75 million (Rs 360 crore) by inundation. Goa could lose 4.32% of its land.
- Rice yield in India could fall by 15-42% and wheat yields by 3-4%. Net agricultural revenues would decline by 12.3% if the temperatures change by 2°C and rainfall by just 7%.
- Fall in production alone could push GDP down by 1.8%-3.4%. If total impact of climate change is considered then as much as 9% of GDP of developing countries like India could be wiped out.
- 5218 glaciers in Himalayas would be impacted. Gangotri glaciers have reduced by almost one-third km in just 13 years. Pindari glacier retreats at rate of 13 m a year already. Gangotri glacier is receding at an annual rate of 30m.
- India stands to lose 125 million tonnes (or 18%) of its rainfed cereal production.

Source: The Times of India, New Delhi, Feb.2007.

5.5.6. Effects on Coral Reefs

The coral reefs are found in the nutrient poor tropical oceans, they are among the most productive ecosystems and their varied habitats support a high diversity of colourful invertebrates and vertebrate life. They occupy 15 to 30% of the continental shelf area. Coral reefs are threatened systems because of fishing practices, pollution, increased disease outbreaks and invasive species, and coral bleaching. Temperature increase, sea level rise and changes in sea water chemistry will have drastic effects on coral reefs.

The increase in CO_2 concentration in the atmosphere may also impact the biota by affecting the marine biogenic calcification. A doubling of CO_2 is expected to cause a 20 to 40% reduction in biogenic calcification of the predominant calcifying organisms including corals.

5.5.7. Spread of Tropical Diseases

World scientific community are getting increasingly alarmed over global warming's impact on human health. Warming climate is creating conditions for spread of infectious diseases, putting millions at risk. Spread by mosquitoes and other insects, illnesses like malaria, dengue, yellow fever and encephalitis are becoming more widespread. Warmer temperature allows these insects to move to other parts of the subcontinent and multiply.

5.6. GLOBAL CLIMATE CHANGE STUDY

5.6.1. Palaeoclimate Records

Climates of the past or palaeoclimates have been a subject matter of research by meteorologists and geographers for the past many decades. The palaeoclimate records give important clues about the climate changes in the past due to natural phenomenon. A wide variety of archives such as marine and lake sediments, ice sheet and cores, tree rings, long-lived corals and archaeological remains give useful information about the past climate changes. Sediments and organic materials can be used to evaluate the past climate, nature and extent of changes. One of the most interesting cases is the examination of glacial ice. Snow changes to glacial ice after recrystalization and increase in the density. The process also traps air bubbles that are analyzed to give information about the concentration of CO_2 at the time of ice formation. Ice sheets are formed layer by layer from the snowfall of each year; with time, the snow is compressed into ice often filled with bubbles of trapped air. The glacial ice is a time capsule that provides information about the atmospheric conditions in the past. The 420,000 year Vostok ice core record of atmospheric CO_2, methane concentration and global temperature through four glacial-interglacial cycles has clearly demonstrated a regular pattern and a tight coupling of their temporal dynamics throughout the record (Singh *et al.*, 2006).

Trees record climate change. Each year a tree forms a new layer of tissue. The width of each year's ring reflects growing conditions such as moisture or sunlight. A wide ring indicates faster growth, whereas a narrow ring indicates slower growth. When the tree is cut, or a core is taken out, these rings can be read like a diary of the climate.

5.6.2. General Circulation Model

The climate models describe the climate situations in numerical terms. The General Circulation Models have been used to predict the future consequences of climate warming on biosphere and its life support system. The GCM's predict atmospheric changes by using variables, such as temperature, relative humidity and wind conditions on a global scale (IPCC, 2001). But there are certain limitations of the computer modelling too.

5.6.3. Remote Sensing and Geographical Information Systems

NASA launched the EOS (Earth Observing System) Terra mission in December 1999, a new satellite series, to describe, understand and simulate the functioning of the earth system (Ranson & Wickland, 2001). This mission holds a great promise to understand and simulate the functioning of the earth system. Geographical Information System (GIS) technology is becoming an essential tool in the effort to understand the process of global change.

5.7. APPROACHES FOR GLOBAL WARMING

In order to stabilize atmospheric concentrations of greenhouse gases, reduction in man-made greenhouse gas emissions is necessary. The various mitigation strategies that could help to reduce the risk of global warming include reduction of energy use by developing more efficient systems, reducing use of fossil fuel, increasing forest cover, minimizing the use of chemical fertilizer for reducing N_2O and improving rice cropping practices for reducing CH_4 emission. Low or no tillage can increase carbon sequestration in the soil. It is known that the amount of carbon in the soils of terrestrial ecosystems is approximately three times that of the atmosphere. The amount of carbon is also 700 times more than the estimated annual increase of CO_2 in the atmosphere. Carbon sequestration in terrestrial ecosystems can occur in living aboveground biomass, and living biomass in soils. For effective carbon sequestration, the increased photosynthetic carbon fixation must occur in the long lived pools. Plants take up CO_2 during photosynthesis and can store carbon in the roots and stems for longer periods of time. It is estimated that about 10 to 15% of the excess carbon dioxide in the atmosphere can be removed by creating large scale tree plantation. In agroecosystems, conservation agriculture, improved varieties of crop plants, increase in plant cover on agricultural soils and improving soil fertility can contribute appreciable to carbon sequestration in agricultural soils.

Carbon trade related to making a provision for the payment to the industries, community and others for undertaking activities that sequester carbon through plantation and conserving forest. It is another approach to mitigating the problem of rise in atmospheric CO_2. Several nations have imposed taxes on greenhouse gas emissions on input basis. According to European Commission, 50% tax is imposed on energy production and 50% tax on carbon emission. Some European countries have carbon based taxes on all types of fossil fuels like oil, gas and coal.

5.8. CLIMATE CHANGE MITIGATION: GLOBAL ACTION

Given the broad consensus among the scientists that global warming and consequent climate change will progressively increase, many international agencies and groups, multi-national corporations, industries, NGOs, and individuals have pooled their resources to avert it through research-cum-action. It is now realised by the international organizations that the atmospheric concentrations of greenhouse gases be significantly lowered than today. Negative impacts from global warming are significant and appropriate action is needed. International climatic policies are fairly a recent phenomenon.

In June 1989, the world conference on changing atmosphere in Toronto led to the formation of Intergovernmental Panel on Climate Change (IPCC), which was jointly established by United Nations Environmental Programme (UNEP) and the World Meteorological Organization. IPCC is a consortium of over 2000 scientists and policy analysts from over 100 nations and has produced four major reports dealing with the various aspects of the Climate change. The purpose of the IPCC was to provide the international community with technological guidance to deal with the problems of climate change.

The United Nations Conference on Environment and Development (UNCED, Earth Summit), held at Rio de Janeiro, Brazil in 1992, established the principles for reducing greenhouse gas emission and adopted the framework convention for climate change. The UNFCCC came into force in March 1994 for a full discussion. The objective of the convention is to achieve stabilization of greenhouse gas concentrations in the atmosphere at a level that would prevent dangerous anthropogenic interference with the climate system. The Kyoto Protocol of the United Nations Framework Convention on Climate Change entered into force on 16 February, 2005. The 37 most industrialized countries of

146 nations ratifying the Kyoto Protocol have agreed to reduce their greenhouse gas emissions below 1990 levels during an initial commitment period of 2008 through 2012. The industrialized countries are allowed to achieve some emissions reductions by investing in energy and tree planting projects in developing countries through Clean Development Mechanism. This protocol requires countries to take appropriate measures to reduce their overall greenhouse gas emissions to a level at least 5% below 1990 level by the commitment period 2008-2012 (Bolin, 1998).

Montreal Protocol is a series of international agreements on the reduction and elimination of CFCs and other ozone depleting substances, which followed the Vienna Convention on the Protection of the Ozone Layer adopted in 1985. In 1987, 27 industrialized countries signed the Montreal Protocol, a landmark international agreement to protect the stratosphere ozone by agreeing to limit the production and use of ozone-depleting substances, phasing out of ozone-depleting substances and helping the developing countries to implement use of alternatives to CFCs. Till date 189 countries have signed the Montreal Protocol for their commitment to environmental sustainability. International Ozone Day marks the signing of Montreal Protocol on 16 September since 1995.

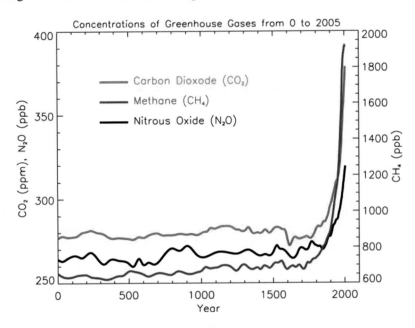

Figure 5.4. Changes in atmospheric concentrations of carbon dioxide, methane, and nitrous oxide from 0 to 2005 years (from IPCC, 2007, reprinted with permission).

5.8.1. India's Initiative for Mitigating Climate Change

India has signed the United Nations Framework Convention on Climate Change (UNFCCC) on 10 June 1992 and ratified it on 1 November 1993. The Ministry of Environment and Forests is the nodal agency for climate change issues in India. It has constituted Working Groups on the UNFCCC and Kyoto Protocol. Current initiatives in India for reducing the greenhouse gas emissions involve the estimation of the sectoral greenhouse gas emissions, national inventory of greenhouse gas sources and sinks, transfer of environmentally sound technology, conservation of forests and biodiversity and coastal zone management besides the involvement of several governmental and non-governmental organizations in climate change research. In the transport sector, India has evolved vehicular emission norm called the 'Bharat 2000", similar to Euro I norm for all vehicles manufactured in the country. It has enforced into the national capital region (NCR), Mumbai, Chennai and Kolkata. Use of the CNG is on the increase.

An expert committee on climate change was constituted by the Ministry of Environment and Forests in 2008. The committee will study the relationship between human activities and global warming, and suggest ways of minimizing their impacts on climate. Carbon management in forests and a proper valuation of ecosystem services could be effective climate change mitigation strategy particularly in the Indian context (Pandey, 2002). The major focus has to be on restoration of degraded lands, expansion of agroforestry, conservation of agriculture and restoration of wastelands for the sequestration of carbon dioxide. Afforestation of wastelands can sequester large amounts of carbon in vegetation and soils.

A national Action Plan on Climate Change was launched in June 2008 to lead India towards sustainable development. The strategies include setting up of 8 national missions on solar energy, enhanced energy efficiency, sustainable habitat, conservation of water, protection of Himalayan ecosystem, greening India, sustainable agriculture and finally establishing a strategic knowledge platform for climate change.

APPENDIX:
CHRONOLOGY OF EVENTS
RELATING TO THE CARBON CYCLE

1750-1820: Industrial revolution. Dramatic increase in the use of coal. Western Europe sees rapid technological, social and economic transformation, driven largely by the steam engine fuelled by coal. Widespread urban pollution, exploitation of workforce, occupational diseases. Human begins to alter the composition of the global atmosphere.

1851: James Young, Scotland, discovers how to extract hydrocarbons from oil shale, and develops the process of refining oil.

1859: Edwin Drake strikes oil at 20 m in Pennsylvania, USA. Oil was soon discovered in North and South America, Mexico, Russia, Iran, Iraq, Rumania, Japan Burma and elsewhere. Oil soon plays its part in the industrialization of the world.

1896: Arrhenius, Swedish chemist, advances theory that carbon dioxide emission will lead to global warming, and postulates the ocean as a global CO_2 sink.

1958: Charles Keeling, of the Scripps Institute in the USA begins the first reliable measurements of atmospheric carbon dioxide at Mauna Loa in Hawaii.

1972: First international conference on the environment, Stockholm, leading to the establishment of the UNEP. Acid Rain was widely publicized, especially in relation to forest decline.

1987: Ice core from Antarctica, taken by French and Russian scientists, reveals close correlation between CO_2 and temperature over the last 100000 years.

1987:   United Nations World Commission on Environment and Development produce the Brundtbland Report, dealing with definitions of sustainability.

1989:   Intergovernmental Panel on Climate Change (IPCC) is established. In June 1989, the world conference on changing atmosphere in Toronto led to the formation of Intergovernmental Panel on Climate Change (IPCC), which was jointly established by UNEP and WMO.IPCC is a consortium of over 2000 scientists and policy analysts from over 100 nations and has produced four major reports dealing with the various aspects of the climate change.

1990:   IPCC's first Scientific Assessment Report, linking greenhouse gas emissions to warming.

1992:   Implementation of the IGBP to predict the effects of changes in climate, atmospheric composition and land use on terrestrial ecosystems; and to determine how these effects lead to feedbacks to the atmosphere.

1992:   UNCED, Earth Summit, Rio de Janerio, Brazil. Leaders of the world's nations meet in Rio, established the principles for reducing greenhouse gas emission and adopted the framework convention for climate change and set out an ambitious agenda to address the environmental, economic, and social challenges facing the international community.

1995:   International Ozone Day.

1997:   Kyoto Protocol, international agreement to limit greenhouse gas emissions.

1997:   Montreal Protocol, international agreements on the reduction and elimination of CFCs and other ozone depleting substances. To date 189 countries have signed the Montreal Protocol for their commitment to environmental sustainability.

1998:   The warmest year of the century, and probably of the millennium.

2001:   President Bush announces that the USA will not ratify the Kyoto Protocol.

2002:   World Summit Sustainable Development.

2007:   The first Climate Change Conference of the Parties to the UNFCCC was held in Bali, Indonesia, wherein a Road Map for future negotiations was laid. It was decided to jointly step up international efforts to combat global warming and climate change and present an agreed outcome in Copenhagen in 2009.

2008:		The Bali conference was followed by the Bangkok meet in March from all the signatories of the Protocol participated in the conference.

REFERENCES

Anonymous (1992) *Action Plan for Himalaya*. HIMAVIKAS Occasional Publication No. 2. G.B. Pant Institute of Himalayan Environment and Development, Kosi, Almora.

Adams, T.M. & Langhuin, R.J. (1981) The effects of agronomy on the carbon and nitrogen contained in soil biomass. *Agriculture Science*, 319-327.

Alford, D. (1992) *Hydrological Aspects of the Himalayan Region*. Occasional Paper No. 18, ICIMOD, Kathmandu, Nepal.

Allen, J.C. & Barnes, D.F. (1985) The causes of deforestation in developing countries. *Annals Association of American Geographers*, 76: 163-184.

BAHC (1993) *Biospheric Aspects of the Hydrological Cycle (BAHC): The Operational Plan*. IGBP Report No 27. BAHC Core Project Office, Berlin.

Bajracharya, S. R., Mool, P.K., & Shrestha, B.R. (2007) *Impact of Climate Change on Himalayan Glaciers and Glacial Lakes*. ICIMOD and UNEP/ROAP, Kathmandu, Nepal.

Bajracharya, S.R., Mool, P.K., & Shrestha B.R. (2006) The impact of global warming on the glaciers of the Himalaya. In: *Proceedings of the International Symposium on Geo-disasters, Infrastructure Management and Protection of World Heritage Sites*, 25-26 November 2006, pp 231-242. Nepal Engineering College, National Society for Earthquake Technology Nepal and Ehime University, Japan, Kathmandu, Nepal.

Baker, R. (1984) Protecting the environment against poor: the historical roots of the soil erosion orthodoxy in the third world. *The Ecologists*, 14 (2): 53-60.

Ball, J.B. (2001) *Global Forest Resources; History and Dynamics*. The Forest Handbook. Oxford, Blackwell Science.

Bandyopadhyay, J. & Gyawali, D. (1994) Himalayan water resources: ecological and political aspects of management. *Mountain Research and Development,* 4 (1): 1-24.

Barnett, T.P., Adam, J.C. & Lettenmaier, D.P. (2005) Potential impacts of a warming climate on water availability in a snow-dominated region. *Nature,* 438 (17): 303-309.

Barnola, J.M., Raynaud, D., Korotkevich, Y.S., & Lorius, C. (1987) Vostok ice core provides 160,000 year record of atmospheric CO_2. *Nature,* 329: 408-414.

Bartarya, S.K. & Valdiya, K.S. (1989) Landslides and erosion in the catchment of Gaula river, Kumaun Lesser Himalaya, India. *Mountain Research and Development,* 9: 405-419.

Batjes, N.H. (1992) Organic matter and carbon dioxide. In: *A Review of Soil Factors and Processes that Control Fluxes of Heat, Moisture and Greenhouse Gases, Vol Technical Paper 23,* (eds. Batjes, N.H. & Bridges, E.M.), pp. 97-148. International Soil References and Information Centre (ISRIC), Wageningen.

Bauer, A., & Black, A.L. (1981) Soil carbon, nitrogen and bulk density comparison in two cropland tillage system after 25 years and in virgin grasslands. *Soil Science Society of America Journal,* 45: 1167-1170.

Becker, A., Avissar, R., Goodrich, D., Moon, D. & Sebruk, B. (eds.) (1994) *Climate-Hydrology-Ecosystem Interrelations in Mountain Regions (CHESMO): An International Initiative for Integrative Research.* IGBP-BAHC/UNEP Workshop, St. Moritz, Switzerland. BAHC Report No.2, Berlin.

Bobba, A. G., Singh, V.P. & Bengtsson, L. (1997) Sustainable development of water resources in India. *Environmental Management,* 21 (3): 367-393.

Bolin, S. (1998) The Kyoto negotiations on climate change: a science perspective. *Science,* 279: 330-331.

Bonan, G.B. (1999) Forest followed the plow: impacts of deforestation on the climate of the United States. *Ecological Application,* 9: 1305-1315.

Bouwman, A.F. (1990) Exchange of greenhouse gases between terrestrial ecosystem and atmosphere. In: *Soils and Greenhouse Effects,* (ed. Bouwman, A.F.), pp. 61-126. John Wiley & Sons, New York.

Bouwman, A.F. (ed.) (1990) *Soils and the Greenhouse Effects.* John Wiley & Sons Ltd., New York.

Bren, L.J. (1980) Hydrology of a small, forested catchment. *Australian Forest Research,* 10: 39-51.

Bren, L.J. & Turner, A.K. (1979) Overland flow on a steep forested infiltrating slope. *Australian Journal of Soil Research,* 17: 43-52.

Burke, I.C., Yonker, C.M., Parton, W.J., Cole, C.V., Flach, K. & Schimel, D.S. (1989) Texture, climate and cultivation effects on soil organic matter content in U.S. grassland soils. *Soil Science Society of America Journal*, 53: 800-805.

Chan, Y.H. (1982) Storage and release of organic carbon in peninsular Malaysia. *International Journal of Environmental Studies*, 18: 211-222.

Chalise, S.R. & Khanal, N.R. (2002) Recent extreme weather events in the Nepal Himalaya. In: *The Extremes of the Extremes: Extraordinary Floods*, Snorrason, A., Finnsdottir, H.P., Moss, M.E. (eds.), pp. 141-146, Proceedings of a Symposium held in Reykjavik, July, 2000. Publication No. 271, Wallingford,.

Chapin, S.F. III., Matson, P.A., & Mooney, H.A. (2002) *Principles of Terrestrial Ecosystem Ecology*. Springer Publishers, New York.

Clarke, W.C. (ed.) (1982) *Carbon Dioxide Review*. Oxford University Press, Oxford.

Cole, C.V., Stewart, J.W.B., Ojima, D.S., Parton, W.J. & Schimel, D.S. (1989). Modelling land-use effects of soil organic matter dynamics in the North American Great Plains. In: *Ecology of Arable Land: Perspectives and Challenges*, Clasholm, M. & Bergstron, L. (eds.), pp.89-98. Kluwer Academic Publishers, Dordrecht, The Netherlands.

Crosby, A.W. (1986) *Ecological Imperialism: The Biological Expansion of Europe, 900-1900*. Cambridge University Press, Cambridge, U.K.

Crutzen, P.J. (1994) Global budget for a non-CO_2 greenhouse gases. *Environmental Monitoring and Assessment,* 31: 1-15.

Crutzen, P.J., & Andreane, M.O. (1990) Biomass in the tropics: impacts of atmospheric chemistry and biogeochemical cycles. *Science,* 250: 1669-1678.

CSE (2002) Melting into Oblivion. *Down To Earth*, 15 May 2002.

Dale, V.H., Brown, S., Haeuber, R.A., Hobbs, N.T., Huntly, N., Naiman, R.J., Riebsame, W.E., Turner, M.G. & Valone, T.J. (2000) Ecological principles and guidelines for managing the use of land. *Ecological Application*, 10: 639-670.

Darby, H.C. (1956) The clearing of the wood-land in Europe. In: *Man's Role in Changing the Face of the Earth,* Thomas, W.L. (ed), University of Chicago Press, Chicago, IL.

Denmead, O.T. (1991) Sources and sinks of greenhouse gases in the soil-plant-environment. *Vegetaio*, 91: 73-86.

Deosthali, V.C. (2000) Impact of rapid urban growth on heat and moisture islands in Pune city, India. *Atmospheric Environment*, 34: 2745-2754.

Dickinson, R.E. & Henderson-Sellers, A. (1988) Modeling tropical deforestation: a study of GCM land-surface parameterizations. *Quarterly Journal of the Royal Meteorological Society*, 114: 439-462.

Dixon, R.K., Brown, S., Houghton, R.A., Soloman, A.M., Trexler, M.C. & Wisniewski, J (1994) Carbon pools and flux of global forest ecosystems. *Science*, 263; 185-190.

Dobhal, D.P., Gergan, J.T., & Thayyen, R.J. (1999) Recession of Dokriani Glacier, Garhwal Himalaya: an overview. In: *Proceedings of Symposium on Snow, Ice and Glaciers: A Himalayan Perspective*, Abstract, pp30-33. Geological Survey of India, Delhi.

Du, M.Y., Kawashima, S., Yonemura, S., Zhang, X.Y., & Chen, S.B. (2004) Mutual influence between human activities and climate change in the Tibetan Plateau during recent years. *Global and Planetary Change*, 41: 241-249.

Eriksson, K. (1991) Sources and sink of carbon dioxide in Sweden. *Ambio*, 20: 146-150.

FAO (1990) *Production Yearbook*. FAO, Rome, Italy.

FAO (1993) *Forest Resources Assessment 1990 Programme: Tropical Countries*. FAO Forestry Paper 112, Rome, Italy.

FAO (2001) *Global Forest Resources Assessment 2000: Main Report*, Rome, Italy.

Feddema, J.J., Oleson, K.W., Bonan, G.B., Mearns, L.O., Buja, L.E., Meehl, G.A., & Washington, W.M. (2005) The importance of land-cover change in simulating future climates. *Science*, 310:1674-1678.

Ferguson, R. (1984) Sediment load of the Hunza river. In: *International Karakorum Project*, Miller, K. (ed.), Cambridge University Press, Cambridge.

Filippelli, G.M. (2008) The global phosphorus cycle: past, present, and future. *Elements*, 4: 89-95.

Frency, J.R., Ivanov, M.V., & Rodhe, H. (1983) The sulphur cycle. Chapter 2, C,N.P. and S cycles: major reservoirs and fluxes. In: *The Major Biogeochemical Cycles and Their Interactions*, Bolin, B., and Cook, R.B. (eds.), pp. 57-60. SCOPE 21, John Wiley, Chichester.

Gaillardet, J., Dupre, B., Louvat, P. & Allegre, C.J. (1999) Global silicate weathering and CO_2 consumption rates deduced from the chemistry of large rivers. *Chemical Geology*, 159:3-30.

Galo, K.P., Easterling, D.R. & Peterson, T.C. (1996)The influence of land use/land cover on climatological values of the diurnal temperature range. *J. Climatology*, 9 (11): 2941-2944.

Global Carbon Project (2003) *Science Framework and Implementation. Earth System Science Partnership.* Reprot No. 1, Canberra.

Global Change Report No. 4 (1988) *The International Geosphere Biosphere Programme: A Study of Global Change.* Royal Swedish Academy of Sciences, Stockholm, Sweeden.

Grace, J. (2004) Understanding and managing the global carbon cycle. *Journal of Ecology*, 92: 189-202.

Hasnain, S.I., Ahmas, S., & Kumar, R. (2004) Impact of Climate Change on Chhota Shigri Glacier, Chenab Basin, Gangotri Glacier, Ganga Headwater in the Himalaya. In: *Proceedings of Workshop on Vulnerability Assessment and Adoption Due to Climate Change on India Water Resources, Coastal Zones and Human Health,* 27-28 June 2003, New Delhi, pp. 1-7. Ministery of Environment and Forests, Government of India.

Hass, H.J., Evance, C.E. & Miles, E.F. (1957) *Nitrogen and Carbon Changes in Great Plains Soils as Influenced by Cropping and Soil treatments.* Technical Bulletin No.1164, USAD, US Government Printing Office, Washington, DC.

Holligan, P.M. & de Boois, H. (1993) *Land-Ocean Interactions in the Coastal Zone (LOICZ). IGBP Report No 25.* International Geospher-Biosphere Programme, Stockholm.

Houghton, J.T. et al., (eds.) (2001) *Climate Change 2001: The Scientific Basis. Contribution of Working Group 1 to the Third Assessment Report of the IPCC.* Cambridge University Press, Cambridge.

Houghton, J.T., Callander, B.A. & Varney, S.K. (eds) (1992) *Climate Change 1992. The Supplementary Report to the IPCC Scientific Assessment.* IPCC, Cambridge University Press, Cambridge.

Houghton, R.A. (1990) The global effects of tropical deforestation. *Environmental Sciences and Technology*, 24, 414-422.

Houghton, R.A. (1994) The worldwide extent of land-use change. *Bioscience*, 44: 305-313.

Houghton, R.A. (1999) The annual net flux of carbon to the atmosphere from changes in land use 1850-1990. *Tellus*, 51B: 298-313.

Houghton, R.A. & Hackler, J.L. (1994) Emissions of carbon from forestry and land-use change in tropical Asia. *Global Change Biology*, 5: 481-492.

Houghton, R.A. & Skole, D.L. (1990) Carbon. In: *The Earth as Transformed by Human Action,* Turner II, B.L., Clark, W.C., Kates, R.W., Richards, J.F., Mathews, J.T. & Mayer, W.B.(eds.), pp. 393-408. Cambridge University Press, Cambridge.

Houghton, R.A., Boone, R.D., Fruci, J.R., Hobbie, J.E., Melillo, J.M., Palm, C.A., Peterson, B.J., Shaver, G.R., Woodwell, G.M., Moore, B., Skole, D.L. &

Myers, N. (1987) The flux of carbon from terrestrial ecosystems to the atmosphere in 1980 due to changes in land-use: geographic distribution of the global flux. *Tellus*, 39B: 122-139.

Houghton, R.A., Boone, R.D., Melillo, J.M., Palm, C.A., Woodwell, G.M., Myers, N., Moore, B. & Skole, D.L. (1985) Net flux of carbon dioxide from terrestrial tropical forests in 1980. *Nature*, 316: 617-620.

Houghton, R.A., Hobbie, J.E., Melillo, J.M., Moore, B., Peterson, B.J., Shaver, G.R. & Wordwell, G.M. (1983) Changes in the carbon content of terrestrial biota and soils between 1860 and 1980: a net release of CO_2 to the atmosphere. *Ecological Monographs*, 53: 235-262.

Houghton, R.A., Lefkowitz, D.S. & Skole, D.L. (1991) Changes in the landscape of Latin America between 1850 and 1985, I: progressive loss of forest. *Forest, Ecology and Management*, 38: 143-172.

Houghton, R.A., Skole, D.L., Nobre, C.A., Hackler, J.L., Lawrence, K.T. & Chomentowski, W.H. (2001) Annual fluxes of carbon from deforestation and regrowth in the Brazilian Amazon. *Nature*, 403: 301-304.

ICDC (2005) *7th International Carbon Dioxide Conference*. Boulder, Colorado, USA, from September 25-30.

IGBP (2001) *Global Change and the Earth System: A Planet under Pressure*. IGBP Science No 4. IGBP Secretariat, Stockholm.

IGBP Terrestrial Carbon Working Groups (1998) The terrestrial carbon cycle implications for the Kyoto protocol. *Science*, 280: 1393-1394.

IPCC (2001) *Land Use, Land-Use Change and Forestry*. Cambridge University Press, Cambridge.

IPCC (2001a) *Climate Change 2001: The Scientific Basis*. Contribution of Working Group I to the Third Assessment Report of the Intergovernmental Panel on Climate Change. Cambridge University Press, Cambridge.

IPCC (2001b) *Climate Change 2001: Impacts, Adaptation and Vulnerability*. Contribution of Working Group II to the Third Assessment Report of the Intergovernmental Panel on Climate Change. Cambrdige University Press, Cambridge.

IPCC (2007) *Climate Change 2007: The Physical Sciences Basis*. In Summary for Policy Makers, IPCC: 21, Geneva.

IUCN; IWMI; Ramsar Convention and WRI (2003) Water Resource Atlas. Available online at http://multimedia. Wri.org/watersheds_2003/index.html, accessed on 12 June 2007.

Ives, J.D., Messerli, B. & Rhoades, R.E. (1997) Agenda for sustainable mountain development. In: *Mountains of the World: A Global Priority*, Messerli, B. & Ives, J.D. (eds.), pp.455-466. Parthenon, Carnforth, U.K.

Jackson, P. (1983) The tragedy of our tropical rain forests. *Ambio,* 12: 252-254.

Jackson, R.B., Banner, J.L., Jobbagy, E.G., Pockman, W.T. & Wall, D.H. (2002) Ecosystem carbon loss with woody plant invasion of grasslands. *Nature,* 418: 623-626.

Jenkinson, D.S., Adams, D.E., & Wild, A. (1991) Model estimates of CO_2 emissions from soils in response to global warming. *Nature:* 351: 304-306.

Johnson, D.W. (1992) Effects of forest management on soil carbon storage. *Water, Air, and Soil Pollution,* 64: 83-120.

Johnson, M.G., & Kern, J.S. (2002) Quantifying the organic carbon held in forested soils of the United States and Puerto Rico. In: *The Potential of U.S. Forest Soils to Sequester Carbon and Mitigate the Greenhouse Effect,* Kimble, J.M., Heath, L.S., Birdsey, R.A., Lal, R. (eds.), pp. 47-72, Lewis Publishers, Boca Raton, FL.

Kale, V.S. & Gupta, A. (2001) *Introduction to Geomorphology.* Orient Longman, Calcutta.

Kalnay, E., & Cai, M. (2003) Impact of urbanization and land-use change on climate. *Nature,* 423: 528-531.

Keeling, C.D. & Whorf, T.P. (2005) Atmospheric CO_2 records from sites in the SIO air sampling network. In: *Trends: A Compendium of Data on Global Change. Carbon Dioxide Information Analysis Center,* Oak Ridge National Laboratory, U.S. Department of Energy, Oak Ridge, Tennessee, USA.

Kellogg, W.W. (1982) Society, Science and Climate Change. *Foreign Aff.,* 60: 1076-1109.

Khalil, M.A.K., & Rasmussen, R.A. (1990) Constraints on the global sources of methane and an analysis on recent budget. *Tellus,* 42B: 229-236.

Khoshoo, T.N. (1992) *Plant Diversity in the Himalaya: Conservation and Utilization.* IInd Pandit Gobind Ballabh Pant Memorial Lecture, Gangtok.

Kim, Y.H. & Baik, J.J. (2002) Maximum urban heat island intensity in Seoul. *Journal of Applied Meteorology,* 41: 651-653.

Kimble, J.M., Heath, L.S., Birdsey, R.A., & Lal, R. (2002) *The Potential of U.S. Forest Soils to Sequester Carbon and Mitigate the Greenhouse Effect.* Lewis Publishers, Boca Raton, FL.

Kotwichi, V. (1991) Water in universe. *Journal of Hydrological Science,* 36: 49-66.

Kulkarni, A.V., Rathore, B.P. & Sujan, A. (2004) Monitoring of glacial mass balance in the Baspa basin using accumulation area ratio method. *Current Science,* 86 (1): 101-106.

Kump, L.R. (2002) Reducing uncertainity about carbon dioxide as a climate driver. *Nature,* 419: 188-190.

Lal, M. (2002) *Possible Impacts of Global Climate Change on Water Availability in India.* Report to Global Environment and Energy in the 21ˢᵗ Century. Indian Institute of Technology, New Delhi.

Lal, P., Kimble, J.M., Follet, R.F., & Stewart, B.A. (2001) *Assessment Methods for Soil Carbon.* Lewis Publishers, Boca Raton, FL.

Laurance, W.F., Cochrane, M.A., & Bergen, S. et al., (2001a) Environment-the future of Brazilian Amazon. *Science,* 291: 438-439.

Laurance, W.F., Fearnside, P.M., & Cochrane, M.A., et al., (2001b) Development of the Brazilian Amazon-Response. *Science,* 292: 1652-1654.

Lauterburg, A. (1993) The Himalayan highland-lowland interactive system: do land use changes in the mountains affect the plains? In: *Himalayan Environment Pressure-Problems-Processes,* Messerli, B., Hofer, T., Wymann, S. (eds.). Geographica Bernensia G38, Institute of Geography, University of Bern.

Likens, G.E., Bormann, F.H., Johnson, N.M., Fisher, D.W. & Pierce, R.S. (1970) Effects of forest cutting and herbicide treatment on nutrient budgets in the Hubbard Brook watershed-ecosystem. *Ecological Monographs,* 40: 23-47.

Liu, X., & Chen, B. (2000) Climatic warming in the Tibetan Plateau during recent decades. *International Journal of Climatology,* 20: 1729-1742.

Loveland, T.R., et al., (2000) Development of a global land cover characteristics database and IGBP discover from 1 km AVHRR data. *International Journal of Remote Sensing,* 21: 1303-1330.

LUCC (1995) *Land-Use and Land-Cover Change: Science/Research Plan.* IGBP Report No 35. IGBP Report No 13. International Geospher-Biosphere Programme, Stockholm.

Malahi, Y., Baldocchi, D.D. & Jarvis, P.G. (1999) The carbon balance of tropical, temperate and boreal forests. *Plant, Cell and Environment,* 22: 715-740.

Mann, L.K. (1986) Changes in soil carbon storage after cultivation. *Soil Science,* 142: 279-288.

Marsh, G.P. (1864) *Man and Nature.* Charles Scribner, New York.

Mather, A.S. (1986) *Land Use.* Longman, London.

Mather, J.R. & Sdasyuk, G.V. (1991) *Global Change: Geographical Approaches.* University of Arizona Press: Tucson.

Matny, L. (2000) Melting of Earth's Ice Cover Reaches New High. In: Worldwatch News Brief, 6 March 2000.

Matson, P.A. & Ojima, D.S. (eds.) (1990) *Terrestrial Biospheric Exchange with Global Atmospheric Chemistry.* IGBP Report No 13. International Geospher-Biosphere Programme, Stockholm.

Mayweski, P. & Jaschke, P.A. (1979) Himalaya and Trans Himalayan glacier fluctuation since A.D. 1812. *Arctic and Alpine Research*, 11 (3): 267-287.

McDowell, N. (2002) Melting ice triggers Himalayan flood warning. *Nature*, 416: 776.

McNeill, J. (2000) *Something New Under the Sun*. Allen Lane & Penguin, London.

Merz, J. (2004) *Water Balances, Floods and Sedimet Transport in the Hindu Kush-Himalaya*. Institute of Geography, University of Berne.

Meybeck, M. & Ragu, A. (1995) *River Discharges to the Oceans: An Assessment of Suspended Solids, Major Ions and Nutrients*. UNEP, Paris.

Meyer, W.B. & Turner, B.L. (1994) *Global Land-use/Land-cover Change*. Cambridge University Press, New York.

Milas, S. (1984) Population crisis and desertification in the Sudano-Sahelian Region. *Environmental Conservation*, 11: 167-169.

Milliman, J.D. & Meade, R.H. (1983) World-wide delivery of river sediments to the oceans. *Journal of Geology*, 91: 1-21.

MOEF (Ministry of Environment and Forests) (2002a) *Empowering People for Sustainable Development*. Ministry of Environment and Forests, Government of India, New Delhi.

MOEF (Ministry of Environment and Forests) (2002b) *Sustainable Development: Learnings and Perspectives from India. Based on Nationwide Consultative* Process. Facilitated by CEE/MOEF/UNDP. Ministry of Environment and Forests, Government of India, New Delhi.

Mool, P.K., Bajracharya, S.R., & Joshi, S.P. (2001) *Inventory of Glaciers, Glacial Lakes and Glacial Lake Outburst Flood Monitoring and Early Warning System in Hindu Kush-Himalayan Region*. ICIMODE, Kathmandu, Nepal.

Myers, N. (1980) *Conversion of Tropical Moist Forests*. National Academy Press, Washington, DC.

Myers, N. (1991) Tropical forests: present status and future outlook. *Climate Change*, 19: 3-32.

NASA Earth Observatory news feature (http://earthobservatory.nasa.gov/study/DeepFreeze/deep_freeze5.html).

NASA news feature, "Tropical Deforestation Affects Rainfall in the U.S. and Around the Globe" (www.nasa.gov/centers/goddard/news/topstory/2005/deforest_rainfall.html).

Nakicenovic, N., Grubler, A., & McDonald, A. (eds.) (1998) *Global Energy Perspectives.* Cambridge University Press, New York.

Naue, N-V. & Roger, P.A. (1993) Rice agriculture: factors controlling emissions, pp 254-298. In: *Atmospheric Methane: Sources, Sinks and Role in Global Change*, Khalil, M.A.K. (ed.). Springer-Verlag, Berlin.

Negi, G.C.S., Rikhari, H.C. & Garkoti, S.C. (1998) The hydrology of three high-altitude forests in Central Himalaya, India: a reconnaissance study. *Hydrological Processes*, 12: 343-350.

Nowak, R.S., Ellsworth, D.S. & Smitch, S.D. (2004) Tansley Review: Functional response of plants to elevated atmospheric CO_2-do photosynthetic and productivity data from FACE experiments support early predictions? *New Phytologist*, 162: 253-280.

Oliver, J.E. & Hidore, J.J. (2003) *Climatology: An Atmospheric Science*, Second Edition. Pearson Education Inc., Delhi, India

Palm, C.A., Houghton, R.A., Mellilo, J.M. & Skole, D.L. (1986) Atmospheric carbon dioxide from deforestation in Southeast Asia. *Biotropica*, 18: 177-188.

Pandey, A.N., Pathak, P.C. & Singh, J.S. (1983) Water, sediment and nutrient movement in forested and non-forested catchments in Kumaun Himalaya. *Forest, Ecology and Management*, 7: 19-29.

Pandey, D.N. (2002) Global climate change and carbon management in multifunctional forests. *Current Science*, 83: 593-602.

Parmesan, C. & Yohe, G. (2003) A globally coherent fingerprint of climate change impacts across natural systems. *Nature*, 421: 37-42.

Pant, R. (1997) Role of traditional institutions in forest management: a case study from Arunachal Pradesh in North-east India. *Arunachal Forest News*, 15: 31-36.

Pendall, E., Bridgham, S., Hanson, P.J., Hungate, B., Kicklighter, D.W., Johnson, D.W., Law, B.E., Luo, Y., Megonigal, J.P., Olsrud, M., Ryan, M.G., & Wan, S. (2004) Below-ground process responses to elevated CO_2 and temperature: A discussion of observations, measurement methods, and models. *New Phytologist*, 162: 311-322.

Petit, J.R., Jouzel, J., Raynaud, D., Barkov, N.I., Barnola, J. –M., Basile, I., Bender, M., Chappellaz, J., Davis, M., Delaygue, G., Delmotte, M., Kotlyakov, V.M., Legrand, M., Lipenkov, V.Y., Lorius, C., Ritz, C., Saltzman, E. & Stievenard, M. (1999) Climate and atmospheric history of the past 420,000 years from the Vostok ice core, Antarctica. *Nature*, 399: 429-436.

Philip, G. & Sah, M.P. (2004) Mapping repeated surges and retreated of glaciers using IRS-1C/1D data: a case study of Shaune Garang glacier, Northwesttern Himalaya. *International Journal of Applied Earth Observation and Geoinformation*, 6 (2): 127-141.

Pielke Sr. R.A. (2005) Land use and climate change. *Science*, 310: 1625-1626.

Pimental, D., Harvey, C., Resosudarmo, P., Sinclair, K., Kurz, D., McNair, M., Crist, S., Shpritz, L., Fitton, L., Saffouri, R., & Blair, R. (1995) Environmental and economic costs of soil erosion and conservation benefits. *Science*, 267 (24): 1117-1123.

Poffenberger, M. (ed.) (2005) *Community Forestry in North-East India.* Community Forestry International, Inc., South Lake Tahoe, USA.

Prasad, V.K., Lata, M. & Badarinath, K.V.S. (2003) Trace gas emissions from biomass burning from Northeast region in India: estimates from satellite remote sensing data and GIS. *The Environmentalist*, 23(3): 229-236.

Prinn, R.G. (1994) The interactive atmosphere: global atmospheric chemistry. *Ambio*, 23: 50-61.

Purohit, A.N. (1991) Potential impact of global climatic change in Himalaya. In: *Impact of Global Climatic Changes on Photosynthesis and Plant Productivity*, Abrol, Y.P., Wattal, P.N., Gnanam, A., Govindjee, Ort, D.R. & Teramura, A.H. (eds.), pp. 591-604. Oxford & IBH Publishing Co. Pvt. Ltd. New Delhi.

Rai, S.C., Sharma, E., & Sundriyal, R.C. (1994) Conservation in the Sikkim Himalaya: traditional knowledge and land-use of the Mamlay watershed. *Environmental Conservation*, 21: 30-34&56.

Rai, S.C. (1995) Land-use options in Sikkim Himalaya: an analysis from the Mamlay watershed of south district. *Journal of Sustainable Agriculture*, 6 (2/3): 63-79

Rai, S.C. & Sharma, E. (1998a) Hydrology and nutrient flux in an agrarian watershed of the Sikkim Himalaya. *Journal of Soil and Water Conservation*, 53: 125-132.

Rai, S.C. & Sharma, E. (1998b) Comparative assessment of runoff characteristics under different land-use patterns within a Himalayan watershed. *Hydrological Processes*, 12: 2235-2248.

Rai, S.C. & Sharma, P. (2004) Carbon flux and land-use/cover change in a Himalayan watershed. *Current Science*, 86 (12): 1594-1596.

Raina, B.N., Hukku, B.M., & Rao, R.V.C. (1980) Geological features of the Himalayan region with special reference to their impact on environmental appreciation and environmental management, pp 1-9. In: *Proceeding National Seminar on Resources, Development and Environment in the Himalayan Region.* Department of Science and Technology, Government of India, New Delhi.

Ramankutty, N., Glodewijk, K.K., Leemans, R., Foley, J. & Oldfield, F. (2001) Land cover change over the last three centuries due to human activities. *Global Change News Letter*, 47: 17-19.

Ranson, K.J. & Wickland, D.E. (2001) EOS Terra: First Data and Mission Status. *Global Change Newsletter*, 45: 23-31.

Rao, K.L. (1979) *India's Water Wealth: Its Assessment, Uses and Projections.* Orient Longman, New Delhi.

Raphael, B.B.M. (1992) Land-use changes and resource degradation in South-West Masailand, Tanzania. *Environmental Conservation*, 19 (2): 145-152.

Ravenstein, E.G. (1890) Lands of the globe still available for European settlement. *Scottish Geographical Magzine*, 6: 541-546.

Rawat, J.S. & Rawat, M.S. (1994) Accelerated erosion and denudation in the Nana Kosi watershed, Central Himalaya, India. Part I: sediment load. *Mountain Research and Development*, 14: 25-38.

Richards, J.F. & Flint, E.P. (1994) *Historic Land use and Carbon Estimates for South and South East Asia: 1880-1980.* Carbon Dioxide Information Analysis Center, Oak Ridge National Laboratory, Numerical Data Package-046, Oak Ridge, T.N.

Richards, J.F. (1990) Land transformation. In: *The Earth as Transformed by Human Action*, (eds. Turner, B.L., Clark, W.C., Kates, R.W., Richards, J.F., Mathews, J.T. & Meyer, W.B.), pp.163-178. Cambridge University Press, Cambridge.

Richey, J.E. (1983) The phosphorus cycle. Chapter 2, C,N.P. and S cycles: major reservoirs and fluxes. In: *The Major Biogeochemical Cycles and Their Interactions*, Bolin, B., and Cook, R.B. (eds.). SCOPE 21, John Wiley, Chichester, pp. 57-60.

Rosswal, T. (1983) The nitrogen cycle. Chapter 2, C,N.P. and S cycles: major reservoirs and fluxes. In: *The Major Biogeochemical Cycles and Their Interactions*, Bolin, B., and Cook, R.B. (eds.), pp. 57-60. SCOPE 21, John Wiley, Chichester.

Rozanov, B.G., Targulian, V. & Orlov, D.S. (1990) Soils. In: *The Earth as Transformed by Human Action,* Turner, B.L., Clark, W.C., Kates, R.W., Richards, J.F., Mathews, J.T. &Meyer, W.B. (eds.), pp. 203-214. Cambridge University Press, Cambridge.

Rubinoff, I. (1983) Strategy for preserving rain forest. *Ambio*, 12: 255-258.

Ruttan, V. (ed.) (1993) *Agriculture, Environment and Health: Towards Sustainable Development into the 21st Century.* University of Minneapolish Press.

Sarin, M.M. (2001) Biogeochemistry of Himalayan rivers as an agent of climate change. *Current Science*, 81 (10): 1446-1450.

Schlesinger, W.H. (1995) An over view of carbon cycle. In: *Soils and Global Change*, Lal, R., Kimble, J., Levine, E., &Stewart, B.A. (eds.). Advance Soil Science, CRC/Lewis Publishers, Boca Raton, FL.

Schlesinger, W.H. (1997) *Biogeochemistry: An Analysis of Global Change*, 2nd edition. Academic Press, San Diego.

Seckler, D., Barker, R., & Amarasinghe, U. (1999) Water scarcity in the twenty-first century. *International Journal of Water Resource Development*, 15: 29-42.

Shah S.L. (1982) Ecological degradation and future of agriculture in the Himalaya. *Indian Journal of Agriculture Economics*, 37: 1-22.

Sharma, E., Rai, S.C. & Sharma, R. (2001) Soil, water and nutrient conservation in mountain farming systems: case-study from the Sikkim Himalaya. *Journal of Environmental Management*, 61: 123-135.

Sharma, P. (2003) *Ecological Linkages of Carbon Dynamics in Relation to Land-use/cover Change in a Himalayan Watershed.* Unpublished Ph. D. Thesis. University of North Bengal.

Sharma, P. & Rai, S.C. (2007) Carbon sequestration with land-use cover change in a Himalayan watershed. *Geoderma,* 139: 371-378

Sharma, P. & Rai, S.C. (2004) Stream flow, sediment and carbon transport from a Himalayan watershed. *Journal of Hydrology*, 289: 190-203.

Shukla, J., Nobre, C. & Sellers, P. (1990) Amazon deforestation and climate change. *Science*, 247: 1322-1325.

Shukla, S.P., & Siddiqui, M.A. (1999) Recession of the Snout Front of Milam Galcier, Goriganga valley, district Pithoragarh, U.P. In: *Proceedings of Symposium on Snow, Ice, and Glaciers: A Himalayan Perspective*, Abstract vol. pp27-29. Geological Survey of India, Delhi.

Singh, J.S., Pandey, A.N. & Pathak, P.C. (1983) A hypothesis to account for the major pathways of soil loss from Himalaya. *Environmental Conservation*, 10 (4): 343-345.

Singh, J.S., Pandey, U. & Tiwari, A.K. (1984) Man and Forests: a Central Himalayan Case study. *Ambio,* 13: 80-87.

Singh, J.S., Singh, L. & Pandey, C.B. (1991) Savannization of dry tropical forest increase carbon flux relative to storage. *Current Science*, 61: 477-479.

Singh, J.S., Singh, S.P. and Gupta, S.R. (2006) *Ecology, Environment and Resource Conservation.* Anamaya Publishers, New Delhi.

Singh, J.S., Tiwari, A.K. & Saxena, A.K. (1985) Himalayan forests: a net source of carbon for the atmosphere. *Environmental Conservation*, 12: 67-69.

Smile, V. (2000) *Cycles of Life*. Scientific American Library, New York.

Southgate, D. (1990) The causes of land degradation along spontaneously expanding agricultural frontiers in the third world. *Land Economics*, 66: 93-101.

Speth, J.G. (1994) *Towards an Effective and Operational International Consumption on Desertification*. International Negotiating Committee, International Convention on Desertification, United Nations, New York.

Srivastava, D. (2003) Recession of Gangotri glacier. In: *Proceedings of Workshop on Gangotri Galcier*, Lukhnow, Abstract, pp 4-6. Geological Survey of India, Delhi.

Steffen, W.L., Walker, B.H., Ingram, J.S. & Koch, G.W. (1992) *Global Change and Terrestrial Ecosystems: The Operational Plan*. IGBP Report No. 21. International Geosphere-Biosphere Programme, Stockholm.

Stoddart, D.R. (1969) World erosion and sedimentation in water. In: *Water, Earth and Man*, R.J. Chorley, (ed.). Methuen, London.

Stone, L. (1990) Conservation and human resources: comments on four case studies from Nepal. *Mountain Research and Development*, 10: 5-6.

Strahler, A., & Strahler A. (2003) *Introducing Physical Geography*. John Wiley & Sons, Inc., USA.

Sudhakar, S., Arrawatia, M.L., Kumar, A., Sengupta, S. & Radhakrishnan, K. (1998) Forest cover mapping of Sikkim: a remote sensing approach. In: *Sikkim: Perspectives for Planning and Development*, Rai, S.C., Sundriyal, R.C. & Sharma, E. (eds.), pp. 205-217. Sikkim Science Society and Bishen Singh Mahendra Pal Singh, Dehradun, India.

Sundriyal, R.C., & Sharma, E. (1996) Anthropogenic pressure on tree structure and biomass in the temperate forests of Mamlay watershed in Sikkim. *Forest, Ecology and Management*, 81: 113-134.

Surendra, K. & Dobhal, D.P. (1994) Snout fluctuation study of Chhota Shigri glacier Lahual and Spiti district, Himachal Pradesh. *Journal Geological Society of India*, 44: 581-585.

Thakur, V.C., Virdi, N.S., Gergan, J.T., Mazari, R.K., Chaujar, R.K., Bartarya, S.K., & Philip, G. (1991) *Report on Gaumukh. The Snout of the Gangotri Galcier*. Unpublished report submitted to Department of Science and Technology, New Delhi.

Thapa, G.B. & Weber, K.E. (1990) Actors and factors of deforestation in tropical Asia. *Environmental Conservation*, 17: 19-27.

Thomas, W.L. Jr. (ed.) (1956) *Man's Role in Changing the Face of the Earth*. University Chicago Press, Chicago.

Toit, R.F. DU (1985) Soil loss, hydrological changes and conservation attitudes, in the Sabi catchment of Zimbabwe. *Environmental Conservation*, 12: 157-166.

Toky, O.P. & Ramakrishnan, P.S. (1981) Run-off and infiltration losses related to shifting agriculture (jhum) in north-eastern India. *Environmental Conservation*, 8: 313-321.

Tucker, R. & Richards, J.F. (1983) *Global Deforestation and the Nineteenth Century World Economy*. Duke University Press, Durham, N.C.

Turner, B.L., Clark, W.C., Kates, R.W., Richards, J.F., Mathews, J.T. & Meyer, W.B. (eds.) (1990) *The Earth as Transformed by Human Action: Global and Regional Changes in the Biosphere over the Past 300 Years*. Cambridge University Press, Cambridge.

Turner, B.L., Meyer, W.B. & Skole, D.L. (1994) Global land-use/land-cover change: towards an integrated program of study. *Ambio*, 23: 91-95.

Turner, B.L., Moss, R.H. & Skole, D. (1993) *Relating Land Use and Global Land Cover Change: A Proposal for an IGBP-HDP Core Projects*. International Geospher-Biospher Programme, Stockholm, Sweden.

Turner, M. (1990) Spatial and temporal analysis of landscape patterns. *Landscape Ecology*, 4: 21-30.

Valdiya, K.S. & Bartarya, S.K. (1991) Hydrogeological studies of springs in the catchment of the Gaula river, Kumaun Lesser Himalaya, India. *Mountain Research and Development*, 11: 239-258.

Vitousek, P.M., Mooney, H.A., Lubchenco, J. & Melillo, J.M. (1997) Human domination of Earth's ecosystems. *Science*, 277: 494-499.

Vohra, C.P. (1981) Note on Recent Glaciological Expedition in Himachal Pradesh. In: Geological Survey of India Special Publication 6, pp 26-29. Geological Survey of India, Delhi.

Wagai, R., Brye, K.R., Grower, S.T., Norman, J.M. & Bundy, L.G. (1998) Land use and environmental factors influencing soil surface CO_2 flux and microbial biomass in natural and managed ecosystems in southern Wisconsin. *Soil Biology and Biochemistry*, 30: 1501-1509.

Walker, S.J., Weiss, R.F. & Salameh, P.K. (2000) Reconstructed histories of the annual mean atmospheric mole fractions for the halocarbons CFC-11, CFC-12, CFC-113, and carbon tetrachloride. *Journal of Geophysical Research*, 105: 14285-14296.

Watson, R.T., Rodhe, H., Oeschger, H. & Siegenthaler, U. (1990) Greenhouse gases and aerosol. In: *Climate Change: The IPCC Scientific Assessment*, Houghton, J.T., Jenkins, G.J. and Ephraums, J.J. (eds.), pp 5-34. Cambridge University Press, Cambridge.

Whitby, M.C. (ed.) (1992) *Land Use Change: The Causes and Consequences.* HMSO, London (for NERC/ITE).

Whitlow, R. (1988) Soil erosion and conservation policy in Zimbabwe: past, present and future. *Land Use Policy*, 5: 419-433.

Williams, M. (1990) Agricultural impacts in temperate wetland. In: *Wetlands: A Threatened Landscape*, Williams, M. (ed.), pp. 181-216. Basil Blackwell, Oxford.

Williams, M. (1994) Forests and tree cover. In: *Changes in Land Use and Land Cover: A Global Perspectives*, Meyer, W.B. & Turner, B.L. (eds.). Cambridge University Press, New York

Wolman, M.G. & Fournier, F.G.A. (ed.) (1987) *Land Transformation in Agriculture.* SCOPE, 32. John Wiley, New York.

Woodwell, G.M., Whittaker, R.H., Reiners, W.A., Likens, G.E., Delwiche, C.C., & Botkin, D.B. (1978) The biota and the world carbon budget. *Science,* 199: 141-146.

WWF (2005) *An Overview of Glaciers, Glaciers Retreat and Subsequent Impacts in Nepal, India and China.* WWF Nepal Programme, Kathmandu.

Yao, T.D., Guo, X.J., Lonnie, T., Duan, K.Q., Wang, N.L., Pu, J.C., Xu, B.Q., Yang, X.X., & Sun, W.Z. (2006)180 record and temperature change over the past 100 years in ice cores on the Tibetan Plateau. *Science in China: Series D Earth Science*, 49 (1): 1-9.

Zheng, S., Fu, C., Xu, X., Huang, Y., Hans, S., Hu, F., & Chen, G. (2000) The Asian nitrogen cycle case study. *Ambio* 31: 79-87.

Zon, R. & Sparthawk, W.N. (1923) *Forest Resources of the World.* McGraw-Hill, New York.

INDEX

D

G